W9-BOI-880

Praise for

INVENTING OURSELVES

'This is a fascinating book, which captures the complexity of adolescence but at the same time provides a clear, accessible account of our current understanding of the "teenage brain". There is much still to learn about how the brain works but for now this book is not only an excellent introduction for someone new to this field but also challenges us all to think again about the importance of the "teenage years" and how we might work with (rather than against) the adolescents with whom we come into contact every day.'

Professor Derek Bell, Director of Learnus

'A very readable book bringing together the up-to-date research about how the adolescent brain develops. This matters to both adolescents and parents but also should be read by everyone who looks after adolescents, be they teachers, doctors or psychologists.'

Professor Dame Sally Davies, Chief Medical Officer for England and Chief Medical Advisor to the UK government

'A brilliant scientist reveals herself to be a writer of considerable talent. With precision, honesty and humour, Sarah-Jayne Blakemore shows how the brooding monsters in your living room are not children gone wrong: they are delicate machines in transition.'

Professor Charles Fernyhough, Professor of Psychology, Durham University

'A superbly engaging account of the development and malleability of the human brain. This is essential reading for educationalists – and indeed for all those interested in how young people's brains develop, and the complex interplay between the environment and the human body.'

Professor Becky Francis, Director of the UCL Institute of Education

'The teenage brain is different, but in what way? This beautifully written book tells just how it influences and is influenced by the new challenging demands of a transformational phase of life. There is no sensationalism here. Sarah-Jayne Blakemore is a pioneer in the field and provides a meticulous account of what we know.'
Dame Professor Uta Frith, Emeritus Professor of Cognitive Development at University College London, and Professor Chris Frith, Professor of Neuropsychology at the Interacting Minds Centre, University of London

'There are few scientists who can write about complex issues with such lucidity and force. Blakemore takes on the topic of adolescence and succeeds and sets a new bar for a no-nonsense discussion about the crucial importance of brain development. Clearly, teenagers are assembling their minds right in front of our very eyes and Blakemore shows us how to understand the wonder and joy of it all.'
Professor Michael Gazzaniga, Director of the SAGE Centre for the Study of the Mind at the University of California Santa Barbara

'I currently live with not one but two adolescent brains and, thanks to the thoughtful, lucid and illuminating scientific insights of Sarah-Jayne Blakemore, now at least I understand why they are so infuriating – and am even encouraged to try to do more to support my kids during this crucial, dramatic and maddening period of brain growth.'
Dr Roger Highfield, Director of External Affairs, Science Museum, London

'In an engaging work of scientific analysis combined with personal anecdote, Professor Blakemore has made an extremely important contribution to the way in which society (and criminal justice in particular) should approach adolescent crime, in particular, gang or group related. The book is thought provoking and should be essential reading for all those considering this difficult issue.'
Sir Brian Leveson, President of the Queen's Bench Division and Head of Criminal Justice

'*Inventing Ourselves* is a gripping celebration of the teenage brain. Essential reading for parents, teachers and teens. Sane, wise, myth-busting, this book is a triumph and should be read by every parent and teacher, but they should be warned. They'll have to fight their teenagers to get this gripping book out of their hands.'
Dr Vivienne Parry, OBE, journalist and author, Head of Engagement at Genomics England

'An irresistible insight into the teenage years. This book is utterly brilliant. Rarely have I read a book on education with such enjoyment.'
Dame Alison Peacock, Chief Executive, Chartered College of Teaching

'Finally, a book about the adolescent brain written by someone who actually does the science! In this highly readable, groundbreaking book Blakemore takes us not only into the minds of teenagers, but into the minds of the people who study them. It's a must-read for anyone interested in how adolescents think, and for everyone concerned about how to apply this knowledge to policy and practice.'
Professor Laurence Steinberg, Laura H. Carnell Professor of Psychology, Temple University, Philadelphia, author of Age of Opportunity

'Beautifully written with clarity, expertise and honesty about the most important subject for all of us. I couldn't put it down.'
Professor Lord Robert Winston, Professor of Science and Society, Imperial College, London

INVENTING OURSELVES

The Secret Life of the Teenage Brain

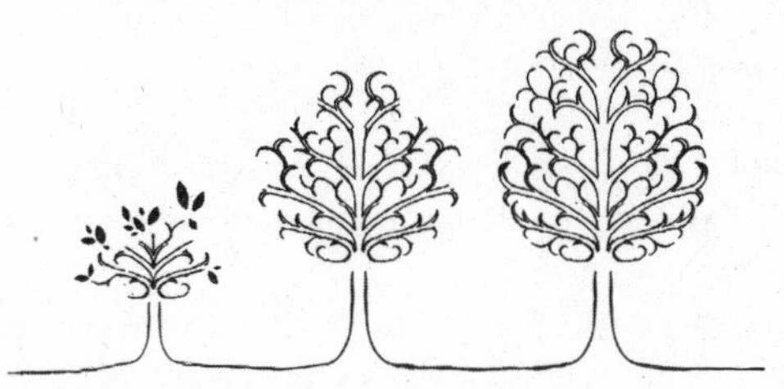

Sarah-Jayne Blakemore

PUBLICAFFAIRS
New York

PublicAffairs
Hachette Book Group
1290 Avenue of the Americas, New York, NY 10104
www.publicaffairsbooks.com
@Public_Affairs

Printed in the United States of America
First published in Great Britain in 2018 by Doubleday, an imprint of Transworld Publishers
First US Edition: May 2018

Published by PublicAffairs, an imprint of Perseus Books, LLC, a subsidiary of Hachette Book Group, Inc. The PublicAffairs name and logo is a trademark of the Hachette Book Group.

The publisher is not responsible for websites (or their content) that are not owned by the publisher.

Typeset in Minion Pro by Falcon Oast Graphic Art Ltd.

ISBNs: 978-1-61039-731-5 (hardcover), 978-1-61039-732-2 (ebook),
978-1-54916-863-5 (audiobook)

Library of Congress Control Number: 2017961853

LSC-C

10 9 8 7 6 5 4 3 2 1

For Oscar and Charlie

Contents

1

Adolescence isn't an aberration

WHEN I TELL PEOPLE I study the adolescent brain, the immediate response is often a joke – something along the lines of: 'What? *Teenagers have brains?*' For some reason it's socially acceptable to mock people in this stage of their lives. But when you think about it, this is strange: we wouldn't ridicule other age groups in the same way. Imagine if we went around openly sneering at the elderly for their poor memory and lack of agility.

Perhaps part of the reason why adolescents are mocked is that they do sometimes behave differently from adults. Some take risks. Many become self-conscious. They go to bed late, get up late. They relate to their friends differently.

The author, aged 15

We now know that all these characteristics are reflections of an important stage of brain development. Adolescence isn't an aberration; it's a crucial stage of our becoming individual and social human beings. I find teenage behaviour fascinating, but not because it's irrational, inexplicable – quite the opposite: because it gives us an insight into how natural changes in the physiology of our brains are reflected in the things we do, and determine who we will become as adults.

In this book, I want to tell you what we know about the adolescent brain. I will show you how we study the way the brain develops during these years, how that development shapes adolescent behaviour, and how it ultimately goes on to define the people we become. This is the time during which much of our sense of ourselves, and of how we fit in with others, is laid down. The development that adolescents go through is central to human experience.

So what *is* adolescence? It's not a straightforward question to answer. Some people think of adolescence as equivalent to the teenage years. Scientific studies often define it simply as the second decade of life – this is the World Health Organization definition. On the other hand, many people believe that adolescence should not be tied to a particular chronological age range. The first psychologist to study adolescence as a period of development was Stanley Hall, who at the beginning of the twentieth century defined adolescence as starting at puberty, around 12 or 13 years, and ending between 22 and 25 years. Many researchers today define adolescence as the interval between the biological changes of puberty and the point at which an individual attains a stable, independent role in society.* In this definition, the start of adolescence is measured biologically while the end is described socially, and is rather arbitrary. In many industrialized cultures the

* The labels given to age groups differ between scientific papers. For example, while some studies classify 13–15-year-olds as 'young adolescents', the same age group might be referred to as 'mid-adolescents' in other studies. Throughout this book I use the labels applied in the original studies and include the age range of each group.

end of adolescence, defined in this way, is constantly being extended as it has become acceptable for young people to stay in full-time education, and live with their parents, into their twenties or even later. Thus, adolescence in the West is often defined as beginning at puberty, now roughly around age 11 or 12, and ending at some point between the late teens and the mid-twenties. In other cultures, things are very different, and children are expected to become financially and socially independent as soon as they reach puberty. In some of these cultures, adolescence isn't seen as a period of development and there's no word for it. Indeed, people often ask whether the concept of adolescence is a recent, Western invention. But it isn't.

There are three main reasons why we can confidently say that adolescence is an important, distinct biological period of development in its own right, in all cultures. First, you can see behaviours that we typically associate with adolescence, such as risk-taking, self-consciousness and peer influence, in many different human cultures, not just those in the West.

A study led by Laurence Steinberg from Temple University in Pennsylvania, and involving scientists from around the world, investigated sensation-seeking and self-regulation in more than five thousand young people from eleven different countries (China, Colombia, Cyprus, India, Italy, Jordan, Kenya, the Philippines, Sweden, Thailand and the United States). Participants aged between 10 and 30 years completed a number of experimental tasks and filled in questionnaires. Two tasks were combined with a questionnaire to provide a measure of *sensation-seeking*, the desire to seek out novel experiences, which often involves risk-taking. A measure of *self-regulation* was also taken – that is, the ability to control yourself and make decisions. Not all cultures showed identical developmental trajectories, but there was remarkable similarity across them. Sensation-seeking increased between age 10 and the late teens (peaking at age 19), and then fell again during the twenties. In contrast, self-regulation increased steadily between 10 and the mid-twenties, after which it levelled out. So, while societal expectations differ between cultures, adolescent-typical behaviours can be seen across cultures.

The second reason why we can consider adolescence a unique period of biological development is that there is also evidence of adolescent-typical behaviour in non-human animals. All mammals undergo a period of development between puberty and becoming fully sexually mature that we can think of as adolescence. There's a lot of research on this period in mice and rats, which are 'adolescent' for about thirty days. Research has shown that, during the month or so of adolescence, these animals take more risks and are more inclined to seek out novel environments than either before puberty or in adulthood. A study published in 2014, carried out by Steinberg and his colleagues, showed that, if given access to alcohol, adolescent mice drink more of it when they are with other adolescent mice; this isn't the case for adult mice.

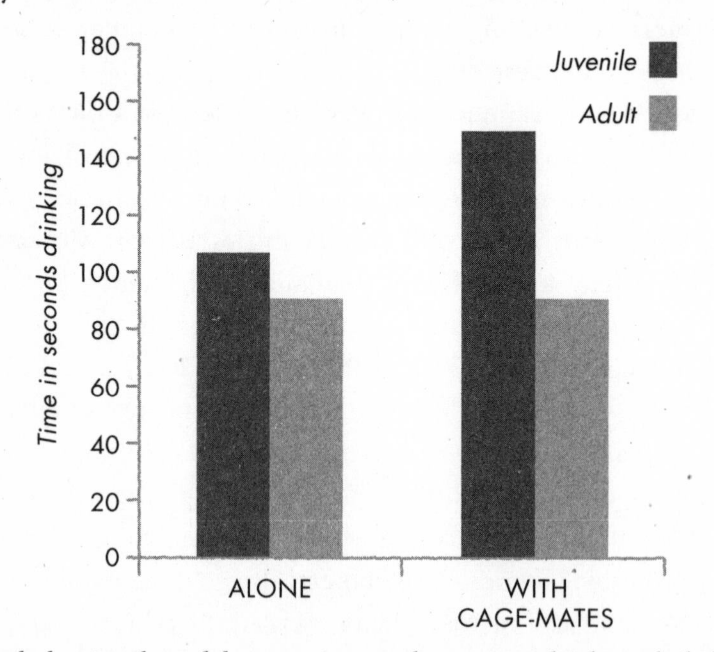

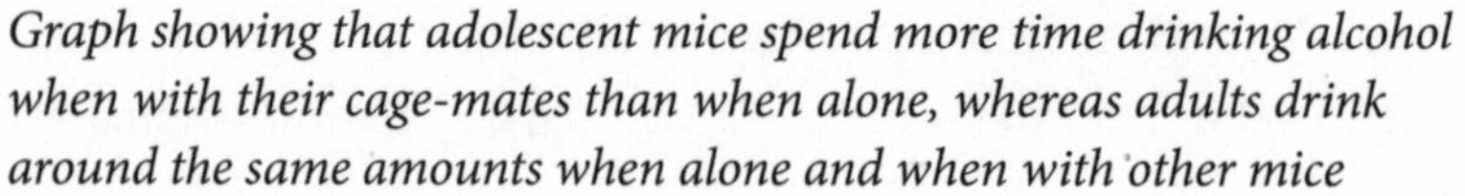
Graph showing that adolescent mice spend more time drinking alcohol when with their cage-mates than when alone, whereas adults drink around the same amounts when alone and when with other mice

We come across adolescent-like behaviour in animals in all sorts of settings. A newspaper article in August 2016 described an incident in which a woman was attacked by an adolescent wombat. In an interview with the *Guardian*, Martin Lind, from the Australian wildlife service, had this to say about the creatures:

> As babies, they're clingy, they're adorable, they're with mum 24 hours a day, they're in a soft, snuggly sleeping bag all the time listening to a heart beat. When they start to mature and hit puberty, they just hate everybody and everything. They go from running between your legs and cute as a button to being absolute little – can I swear? – little shits. They nip you, they wreck, they bite. I won't look after wombats because you kiss goodbye to your flooring and everything. They just destroy everything.

So, adolescent-typical behaviour is present across human cultures and across species. And, third, such behaviour is also typical across *history*. One of the earliest descriptions of adolescents I'm aware of is said to come from Socrates (469–399 BC): 'The children now love luxury. They have bad manners, contempt for authority; they show disrespect for elders and love chatter in place of exercise.' A hundred years or so later, Aristotle described 'youth' as 'lacking in sexual self-restraint, fickle in their desires, passionate and impulsive.'*

* It's worth reading the full passage by Aristotle: 'The young are in character prone to desire and ready to carry any meaningful desire they may have formed into action . . . They are changeful, too, and fickle in their desires, which are as transitory as they are vehement; for their wishes are keen without being permanent . . . they are passionate, irascible, and apt to be carried away by their impulses . . . Youth is the age when people are most devoted to their friends or relations or companions, as they are then extremely fond of social intercourse and have not yet learned to judge their friends, or indeed anything else, by the rule of expediency. If the young commit a fault, it is always on the side of excess and exaggeration; for they carry everything too far, whether it be their love or hatred or anything else. They regard themselves as omniscient and positive in their assertions; this is in fact the reason of their carrying everything too far . . . their offences take the line of insolence and not of meanness.'

Move on a thousand years and more, and we find the Shepherd in Shakespeare's *The Winter's Tale* (1611) complaining (to knowing laughter in the stalls): 'I would there were no age between ten and three-and-twenty, or that youth would sleep out the rest; for there is nothing in the between but getting wenches with child, wronging the ancientry, stealing, fighting.' A century after that, differentiating adolescents from children, Jean-Jacques Rousseau (1712–78) described adolescence thus: 'A change in humour, frequent anger, a mind in constant agitation makes the child almost unmanageable. His feverishness turns him into a lion. He disregards his guide; he no longer wishes to be governed.'

For millennia, then, adolescents have been painted in the same stereotypical terms often used of them today. This suggests that adolescent-typical behaviour is not a recent, Western phenomenon.

What is it that makes adolescents behave in a recognizably 'adolescent' way? Adolescents have long been blamed for their apparently errant ways; some have put their behaviour down to changes in hormones at puberty; others attribute it to social changes following on from puberty and the new importance of peer relationships, or associated with the shift from small primary schools to large secondary schools in early adolescence. Now, though, armed with new knowledge from brain scans and experimental studies, we can try to understand adolescent-typical behaviour in terms of the underlying changes in the brain that happen during these years. Studying changes in brain structure and function reveals a huge amount about why teenagers do what they do, and more broadly about how the architecture of the brain relates to the behaviour we display, and how brain development – as well as hormones and the social environment – shapes who we become as we emerge into adulthood.

There is more at stake here than the advancement of scientific knowledge. Understanding brain development in adolescence has profound implications for social and education policy. Public health advertising aimed at young people, for example, often focuses on the long-term health outcomes of risky activities such as smoking. But there is evidence that this is unlikely to work. Adolescents aren't stupid – rationally, they already understand the risks. But in the heat of the moment, when they're offered a cigarette or an Ecstasy tablet, many adolescents care far more about what their peer group thinks of them than about the potential health risks of their choice. Often, their decisions are driven by the fear of exclusion by their friends, rather than by a dispassionate consideration of the consequences. This isn't true for all adolescents – some young people are not particularly influenced by what their friends think or do – but many are.

Having said this, the new studies on the adolescent brain are also fascinating for their own sake. Contrary to the received wisdom up to the late twentieth century, we now know that our brains are dynamic and constantly changing into adulthood, and that the transformation they undergo in early life continues for far longer and has much bigger implications than was previously thought. Modern brain-scanning technology like magnetic resonance imaging (MRI) is ushering in a new era of understanding of the physiological mechanisms that underpin our sense of who we are, the sense of self that develops during adolescence. In the chapters that follow, I will describe how these technological developments, building on earlier ground-breaking work done with painstaking dexterity on brain tissue samples under microscopes, have revealed the changes that take place in the adolescent brain.

We shouldn't demonize adolescence – it is fundamental to who we are. The adolescent brain isn't a dysfunctional or a defective adult brain. Adolescence is a formative period of life, when neural pathways are malleable, and passion and creativity run high. The changes that take place in the brain during this period offer us a lens through which we can begin to see ourselves anew.

∞

The moment I decided that I wanted to spend my life studying the human brain was the moment I first held one in my hands. I was studying experimental psychology at the University of Oxford. The first year of the degree was split between courses in psychology and neurophysiology, and the latter involved learning about the anatomy of the brain. It included a neuroanatomy practical course during which we were able to study and dissect a human brain. I needed a lab coat and I didn't have one, so I borrowed one from a friend of mine, who was doing a medical degree. Wearing the lab coat, aware that it gave me a new and different identity, I walked briskly into the Anatomy Department and into the large central hall, where students have been dissecting bodies for centuries. I was met with an overwhelming stench of formaldehyde, the liquid used to preserve and store body parts.

In front of each one of us seated around the benches was a large white bucket, closed with a lid, containing a human brain. After a preamble, the lecturer asked us to put on latex gloves and lift the brain with both hands out of the bucket. At that moment, as I held the brain of a stranger – I presumed it to have been that of an elderly man, though I had no evidence for this – I decided that this 3lb mass of tissue had to be the most fascinating and complex object in the universe. I already knew, intellectually, that this was the case, but now I *felt* it, too. The first thing I noticed was how heavy it was, and it astonished me that we all carry one of these around in our heads. I also noticed the colour – greyish pink; and the texture – smooth and shiny, but with folds all over the surface. Simultaneously, I was struck by the realization that the matter this object was made from was synonymous with the person who owned it. I was holding someone in my hands. Over the course of a lifetime, this now disembodied brain had stored all of their memories, had generated all of their feelings, emotions and desires, had formed their personality, their aspirations and their dreams. All of this, all of you, your self, is contained in your brain. *That* was what I was holding in my hands. And I knew then I had to – somehow – spend my life studying it.

∞

By the time I was an undergraduate, I'd already had some acquaintance with what happens when the brain goes wrong.

When my friend Jon* was young, he was a typical kid, if there is such a thing. That's the way his parents expressed it: he had lots of friends, moved in the same circles as his elder brother and sister. At secondary school, he still had a small circle of friends; he played in a band, went out at the weekend, had a girlfriend and got on with his school work.

When Jon was 16, his elder brother, Ben, dropped out of his first year at university because he'd developed schizophrenia. He was back living at his parents' house, so Jon saw him most evenings. Ben was constantly talking to himself. He hardly ever came out of his room. A few times he flipped out and was violent towards his parents. Jon was scared of him and couldn't believe this was his brother – they used to be so close.

Jon left school after his A-levels. He took a gap year and travelled to Asia. All the while, he kept thinking about Ben. And then one day, sitting in a hotel room in Bangkok, Jon heard a voice talking to him, completely clearly, as though he had tuned into a radio station no one else could hear.

One of my main interests as an undergraduate was schizophrenia. Perhaps my interest stemmed from the knowledge that my school-friend Jon had developed this condition just a few years after his brother Ben was diagnosed with the same illness. Schizophrenia is a

* Names in this section have been changed.

devastating psychiatric disorder in which the patient loses touch with reality. It runs in families, but is not completely genetic – the environment plays a role in triggering the illness in people who have a genetic predisposition, although we don't yet understand how exactly this happens.*

People with schizophrenia often have auditory hallucinations, such as hearing voices – usually negative and threatening voices – inside their head. They also often have delusions (false beliefs), such as being paranoid, believing that people are out to get them. A common false belief is that an intelligence agency like MI5 is following their every move. What interested me most was how the human brain actually generates these often terrifying experiences – what goes wrong? And why are most of us protected from them?

Or *are* we? Maybe 'normality' is a fragile state that can be disrupted by taking a drug or by a particularly stressful life event. When I had fevers as a child, my mind sometimes used to play tricks on me. I hallucinated and heard voices. It was frightening, but what I was suffering from is fairly common. It's called fever-induced delirium: the brain heats up because of the fever, which causes neurons to fire in ways that produce unusual, false perceptions. The brain, it turns out, is a delicate ecosystem: nudge things out of balance, and the entire system can be pushed over the edge into the unusual and sinister. So what pushes it?

It was that question – why some people experience delusions and hallucinations, and why most of us don't – that prompted me to apply to do a PhD on schizophrenia. What is it about how our brains work that means that most of us *don't* hear voices or think that the Secret Service

* Although an individual who has a sibling with schizophrenia is more likely to develop the condition than an individual with no schizophrenic siblings, it's still quite rare that two siblings both have this condition (you can read more about this in Gottesman's 1990 book *Schizophrenia genesis*). The exception is identical twins, who share identical genes – in this case, if one twin has schizophrenia, there's around a 50 per cent chance that the other twin will have it too.

is after us? During my PhD, at University College London (UCL), my supervisors Chris Frith and Daniel Wolpert and I found that the brain has a system for labelling self-produced sensations and distinguishing them from sensations produced externally. It turns out that the mechanism for distinguishing between what the world does to you and what you do to yourself isn't working normally in people with schizophrenia. This is probably the reason why they hear their own thoughts as voices, like Jon did, or believe, for example, that their arm movements are being controlled by someone else. Why do people with schizophrenia hear voices that are often negative and critical? Why do they experience paranoid delusions? Why are people with schizophrenia also often depressed, their emotions and demeanour flattened? These were the kinds of questions I was interested in studying.

∞

During my PhD, I collaborated with psychiatrists in Edinburgh and collected data from patients with schizophrenia in psychiatric hospitals there. When I had completed my doctorate and moved on to my post-doctoral research, I also collected data from patients in a psychiatric hospital in Versailles, just outside Paris. As I worked with these people, I was struck time and again by the same observation. Every patient I asked, regardless of age, race or gender, told me that the first time they experienced their frightening and debilitating symptoms was between the ages of 18 and 25 – that is, in what is generally considered late adolescence or emerging adulthood.

It was always the same story. They were pretty regular children, and as teenagers they were OK – some of them started to drop out or take drugs, but not all. There were variations in how fast the disorder appeared – in some cases it had been quite gradual, whereas in others it had happened quickly – but for all of the patients I talked to, it was in their late adolescence that the symptoms started to emerge. This is

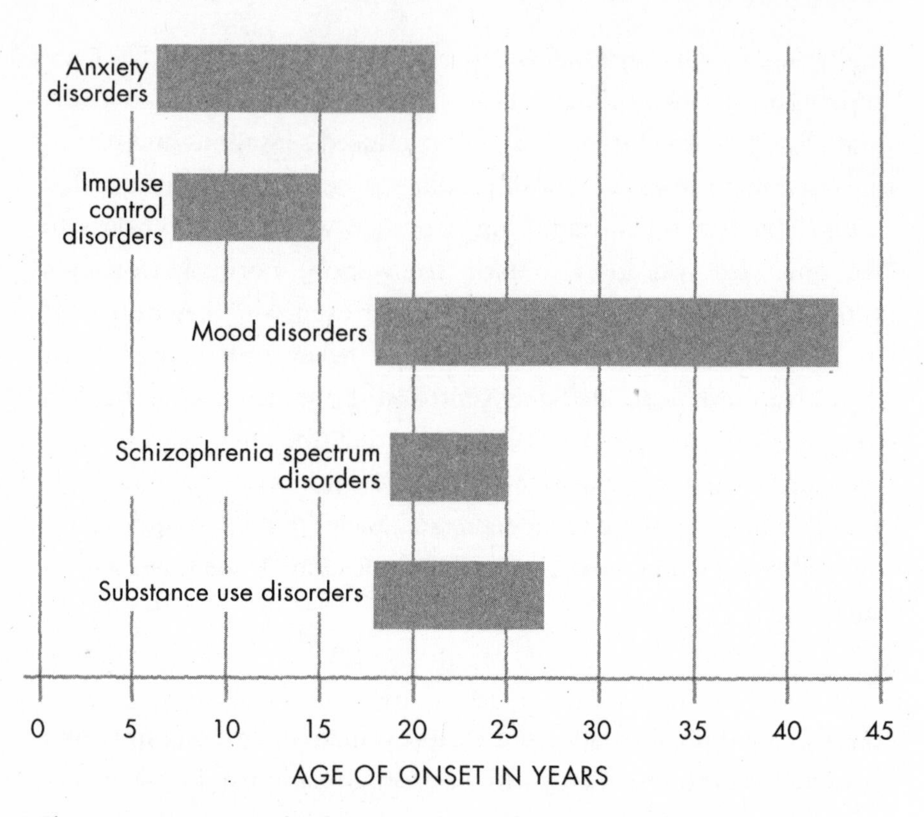

The average age range for the onset of several psychological disorders is in adolescence or early adulthood

interesting because it tells us that schizophrenia is a developmental condition, but one that starts much later than other developmental conditions such as autism or attention deficit hyperactivity disorder (ADHD).

What is it about late adolescence that makes some brains particularly vulnerable to debilitating delusions and hallucinations? What is it about brain development in the teenage years that goes wrong in people who develop schizophrenia? These seemed to me to be critical questions; they also seemed obvious ones to ask, so I thought I would find the answers in the existing scientific literature. But after a concentrated effort delving into journals, to my surprise I discovered how little was known about how even the healthy human teenage brain develops, let alone how it does so in teenagers who go on to develop schizophrenia – and not just schizophrenia: many psychological and psychiatric conditions start at some point in adolescence, as the chart opposite shows. So I became deeply interested in finding the answers for myself.

That was in the year 2001. It was an exciting time to be a researcher. Most neuroscientists still assumed that the human brain doesn't change much after mid- or late childhood. That was what the textbooks said. However, a small handful of studies, published in the late 1990s, suggested that the dogma was completely wrong and that the human brain continues to develop throughout the teenage years, and even into the twenties.

So, after my time in France, I changed the focus of my research from studies of adults with schizophrenia to developmental studies of the teenage human brain. In retrospect, it was a risky manoeuvre, because I'd never done a single developmental study, and there are many aspects of developmental research that are specific to studies of children and adolescents, and to theories of the developing mind, which I had never learned or experienced. Nor did I have experience of the technicalities involving recruiting and testing children and adolescents. It was the encouragement of my friend and mentor Uta Frith which gave me the confidence to make the leap into this new field. Frith is a professor of psychology in London and a world expert on developmental conditions such as autism and dyslexia. I had already known her for several years, having done work experience in her lab when I was 15.* My career jump into adolescent brain development was further aided by a research fellowship from the Royal Society, which started in 2004. I have been working on the adolescent brain ever since.

* My PhD supervisor Chris Frith is Uta's husband. Uta and Chris are still mentors to me, people I reach out to when I have a problem or dilemma.

∞

A window is opened on the adolescent mind by teenage diaries, as vividly demonstrated on the BBC Radio 4 programme *My Teenage Diary*, in which famous people read out their (occasionally painfully embarrassing) diaries from their teenage years.* I recently discovered my own. My parents were clearing out the loft of the house they had been living in for almost forty years and presented me with a large number of musty boxes containing all sorts of things, from old school books to records. One box contained letters I had received as an adolescent – from friends, pen-pals, boyfriends – and the diaries I had written in those years. I'd forgotten that I'd even kept my teenage diaries, and felt a mixture of intrigue and trepidation at the prospect of reading them. As I anticipated, they revealed a fairly typical teenage girl, preoccupied by clothes, music, friends and boys, with occasional heartfelt interjections about the horrors of war (the Gulf War was happening at the time).

So, I found, I was a typical teenager. Perhaps most of us were. Sometimes I wonder whether we forget our own adolescent years when thinking about teenage behaviour today. Adults are quick to criticize teenagers for their moodiness, self-absorption and risky decisions – but, as we've already seen, they have been doing so for over two thousand years at least: Socrates and Aristotle were just as dismissive and critical of young people in ancient Greece as any twenty-first-century parent or teacher. Adolescent-typical behaviour – at least as viewed by adults – goes back a long way.

However, there was another side to my teenage years that distinguished me from my peers. While I spent a lot of time thinking about what clothes to buy from the army surplus store, whether I could afford to go to a gig on Saturday night and who my friends had crushes

* This programme is produced by a good friend of mine, Harriet Jaine.

on, in the background of my life something more serious and threatening was happening. My father, Colin Blakemore, did medical research involving animals, and was targeted by animal rights groups in the UK – in fact, for several years he was their number one target. This meant we lived our lives under the constant threat of attack. The animal rights activists threatened to kidnap me and my two younger sisters, resulting in the three of us, aged between 6 and 11 at the time, being followed to and from school each day by undercover police in an unmarked car. It was a period of my life that left a particularly deep imprint on my mind. For some reason, the police were dressed in leather jackets with piercings, and stuck out like a sore thumb in our quiet, leafy neighbourhood of north Oxford. My sisters and I would walk the few blocks to school, and the police would follow us slowly in their clapped-out car. We found it amusing, but I also remember being intensely embarrassed and hoping none of my friends would spot them.

Our house was under twenty-four-hour guard and was like a fortress, with electric gates, multiple alarm systems, cameras and panic buttons that directly alerted the police. Each time we wanted to drive anywhere, my parents would check under the car with a bomb-detector mirror. Getting into a car that might explode as soon as the ignition was turned on was not an experience I enjoyed much.

The threats and the violence got worse and worse throughout my teenage years. Large numbers of animal rights activists would meet on Saturdays in Oxford, often right outside my parents' house. People would shout abuse through a megaphone and argue vociferously with my dad, who would bravely and calmly confront them. (My 80-year-old grandmother, who lived with us at the time, had a more abrupt approach to the situation, and simply told them to 'bugger off'.)

Sometimes the police got wind of these gatherings beforehand – it's hard to imagine how it was all organized so rapidly and how the police found out, because this was long before everyone had mobile phones and internet access. When this happened, there was a squad car outside the house, and uniformed police with dogs patrolling in the street. One year all the animal rights groups in Britain chose our street as the meeting point for their annual day of protest. Our suburban road was filled with police on horseback, who didn't allow anyone to enter it. Even our neighbours weren't allowed to drive into their own driveways. Again, this was incredibly embarrassing for my teenage self. Interestingly, I didn't really worry too much about the danger of these confrontations, or about the inconvenience for others, not even my parents. I was mostly focused on what other people, and particularly my friends, would think of me: something different was happening to me that didn't happen to anyone else, and when it comes to our teenage lives, 'different' is often bad.

Things got serious. Windows were smashed in our house, bricks thrown through them by a gang wearing balaclavas. Several times our car was doused in paint-stripper, which, aside from the obvious cosmetic damage, also caused chemical burns on the paws of our cat, who liked to sleep on the car. Letters packed with razor blades were sent to my dad through the post.

All this culminated in a bomb being delivered to our house just before Christmas in 1993. I had just started at university, and I was back at my family home after the first term. On 22 December, a parcel that looked like a poster tube was delivered. I picked it up, thinking it was for me, but noticed my dad's name on the front so left it for him under a table near the front door. He got back late, saw the mailing tube – and fortunately remembered the regular warnings from the police about unusual packages. He noticed that it was heavy at one end – not at all like a poster – and decided to leave it in the boot of our car overnight. The police took it away the next morning; later, they told him that it was a lethal bomb, containing half a pound of explosive. If I had opened it, just to take a peek at the contents, I would have been killed.

The following term, that incident was reported in my university newspaper, and I was named in the article. I was furious. The last thing I wanted was any attention drawn to the event or to me – it was highly embarrassing and I was very self-conscious about it. I didn't want to be labelled as a person whose family was targeted by animal rights groups. I wanted to keep it under the radar, for lots of reasons – for one thing, I didn't want to be seen as different, and for another, what if some of my new university friends were opposed to medical research on animals? Would they hold it against me somehow? The article made it all too real, whereas I was in denial, trying to pretend the situation wasn't really as frightening as it was.

Things gradually improved, partly because other scientists started to speak out in support of animal research, partly because the media turned against the shocking tactics of the real extremists, but mostly because of new anti-terrorist legislation. When I think back, even though my parents did their best to shield my sisters and me from the horror of it all, I can't but assume that the whole episode must have had a formative effect on our development – after all, we know that environmental experiences influence brain development during childhood and adolescence. But of course, I have no idea *how* these experiences influenced my development or that of my sisters, for there was no control experiment in which we didn't live in the context of animal rights extremists targeting us.

I wonder how my sense of self was affected. Would I have become a different person had we not had to endure years of threat from the animal rights groups? How did this experience interact with the many other, more typical, experiences of my adolescence, and how did my genetic heritage interact with all those environmental experiences? Would my development have been different in today's world, where I could post my experiences and thoughts on social media? Of course it's impossible to say, but perhaps being surrounded by threat for so many years affected the development of the regions of my brain involved in threat perception and fear processing. Perhaps, as a consequence, I became less (or maybe more) fearful of danger than I would otherwise have become. Of course, any effects on my brain would have been subtle, because my experience of threat was relatively mild. The development of the brain would be much more significantly affected by exposure to severe and constant threat, such as that experienced by children who grow up in war zones or in abusive and chaotic homes.

One way of looking at how experience shapes the adolescent brain is to compare brain and behavioural development in different environments – for example, different cultures. Most of what we know about the adolescent brain and behaviour comes from studies in the United States and Europe. So while we are beginning to learn a lot about Western adolescent brains, we still know very little about brain development in other cultures. It's likely that a variety of environmental experiences, such as the peer group and family environment, education and learning, spending time on the internet or playing video-games, drinking, smoking or taking drugs, play a role in shaping the adolescent brain. We don't yet understand their precise effects, but neuroscience studies are starting to look at how specific environmental factors, including culture, influence brain development. I will come back to this later in the book.

There's a lot we still don't know – yet. We *do* know that adolescence is a formative and protracted period of life during which the sense of self undergoes profound transformation. Younger children do, of course, have a sense of self; a basic one develops very early. But developing an identity is what adolescence is all about. Of course, self-identity is more than just announcing our tastes with clothes, books and posters. During adolescence, your sense of who you are – your moral and political beliefs, your music and fashion tastes, what social group you associate with – undergoes profound change. During adolescence, we are inventing ourselves.

2

A sense of self

HERE'S A BEAUTIFUL ILLUSTRATION of typical adolescent behaviour, from a letter sent by Dinah Hall of Lustleigh, Devon, to the *Guardian* newspaper in 2013:

> There's nothing like teenage diaries for putting momentous historical events in perspective. This is my entry for 20 July 1969.
>
> *I went to arts centre (by myself!) in yellow cords and blouse. Ian was there but he didn't speak to me. Got rhyme put in my handbag from someone who's apparently got a crush on me. It's Nicholas I think. UGH.*
>
> *Man landed on moon.*

Dinah's diary entry illustrates wonderfully what was important to this teenage girl at that particular moment in her life. Evidently, to her, the fact that humans took their first steps on the moon that day was less important than what she was wearing, whom she liked and whom she didn't like. Adolescence is often the first time we give much thought to how our identity affects our lives and the ways in which other people see us. Everything about who we are starts to change. We start to develop a more complex sense of morality and to become aware of the political realities of society around us. We start to develop musical tastes that, for some of us at least, will last a lifetime. Our tastes in fashion also start to develop, although for most of us they don't last as long. We develop more intricate social groups. In short, we are constructing who we are and how we are seen by others.

The first thing I did as a 19-year-old undergraduate at university was unpack my possessions. I hung posters of Bob Dylan and Jimi Hendrix on the walls, set out ornaments I'd acquired from different countries over the past few years, and lined my shelves with books, records and CDs. These possessions represented something about me, *my* self, what was important to me and how I wanted other people to see me. My (largely unread) copies of *Ulysses* and *War and Peace* were proudly displayed (I'm now embarrassed to admit). When had I acquired this sense of self, and how?

A rudimentary sense of self develops very early in life. In the 1990s, Philippe Rochat and his colleagues from Lausanne in Switzerland set up a laboratory in a maternity ward and, with permission from the parents, carried out some simple tests with newborn babies. In an elegant experiment, the researchers compared each baby's responses when his or her cheek was touched either by a researcher or by the baby's own hand (when the baby accidentally brushed his own cheek, for example). The touch was (roughly) the same, but the source of the touch was different. Intriguingly, within twenty-four hours of birth, the babies were able to detect the difference. When someone else touched the baby's cheek, the baby tended to turn his or her head more towards this external touch than they did towards spontaneous self-touching. This simple experiment showed something profound about newborn babies: they were able to distinguish between self and other.

Human babies, then, are born with a sense of self – albeit a very primitive one. Of course, this is very different from our adult sense of who we are and what other people think of us. It's more of an awareness of one's own physiology and the difference between one's own body and those of other people. Even so, it is important, because we need to be able to distinguish between self and other to interact with other people; and it's through interacting with people around them that babies begin to learn language and social behaviour.

After birth, the sense of self develops gradually. By 6 months, babies will look longer at a video of another same-aged infant wearing identical clothes than at a video of themselves. Looking longer suggests not only that they are able to perceive the difference between themselves and other babies, but that they're more interested in the image of the other infant. This is remarkable because it's unlikely that 6-month-olds have spent much time looking at themselves, either in a mirror or on video. Perhaps by 6 months babies have a basic awareness about the way they move, so that the video of themselves has a certain familiarity and is therefore less interesting to look at than the video of another baby moving around.

In the second year of life, self-awareness becomes more explicit: by 18 months, infants develop more advanced signs of an understanding that they exist as individuals, separate from other people. A classic demonstration of this is the mirror self-recognition test. In a series of landmark experiments, researchers applied a coloured mark with a soft brush on some babies' foreheads; on other babies, the researchers touched the forehead with the same brush but didn't apply the mark. What would these two groups of babies do when shown their reflections in a mirror? Before the age of 18 months, babies in both groups paid little attention to their reflections. From around 18 months, the babies with a mark on their foreheads (but not those without a mark) inspected the mark and often tried to rub it off themselves. So, by 18 months, babies seem to have acquired a sense that their bodies and faces belong to them, and they appear to understand that this self is reflected in a mirror. By 2 or 3 years, children start to show an understanding that other people are self-aware just like they are, and they start to differentiate between themselves and other people in speech. The sense of self continues to develop throughout childhood.

How do you become *you*? To a certain degree, our genetic material, inherited from our parents, determines who we will become. Personality, intelligence, preferences and cognitive strengths are all to some degree hereditary – passed on in the genes from generation to generation. In addition, and through interaction with our genes, the environment plays a role in determining who we become. Our many childhood experiences, our upbringing, education, social interactions, hobbies and so on, combine gradually to form a sense of who we are.

It is during adolescence that our sense of self becomes particularly important to us. In my first few weeks of university, aged 19, I was keen to join the student women's group. Women's issues had become important to me during my gap year between school and university. I'd spent three months working in neuroscience laboratories in California where I had been exposed, for the first time, to the social stereotype that it is unusual for a woman to be a scientist. This hadn't been an issue at my all-girls secondary school – none of my teachers had mentioned this stereotype, and many of the pupils studied science for A-level and went on to read scientific subjects at university.

But in California, where I was living independently for the first time, I became aware that many people had preconceptions of what women should be doing with their lives, and that the career I wanted to enter was not viewed as particularly 'feminine'. Scientists were men with white coats and unkempt hair, who spent all day and night toiling away in the lab, weren't they? I spent time with people in California who were vocal about women's equality. These were people I met in the science labs where I was working, and they were an inspiration to me. A professor of psychology at the University of Southern California, Elaine Andersen, who is still a friend and mentor, made a particularly strong impression. She believed that science was not something that only men could or should do, and I agreed with her.

After three months in California I spent six months teaching English in a school near Kathmandu in Nepal. I taught children aged from 3 to 18 years (I was only 18 myself at the time!), and lived with a Nepalese family. Here my eyes were opened to a completely different way of life, and to the vast cultural differences between Nepalese society and my native Britain. When I wasn't teaching, I spent many hours on my own, or with my teaching partner (and now close friend), Camilla, in our small, basic room, reading, writing letters and talking about our experiences at the school.

One topic we discussed a lot was the difference between the societal expectations for men and women in Nepal. At the Hindu school in which I was teaching, there were approximately equal numbers of boys and girls in the primary years until around age 11, after which point most of the girls stopped coming to school. All the children I taught over the age of 13 were boys. It was assumed that the girls would not be educated after childhood; at around the age of puberty, they were expected to give up schooling and instead perform domestic duties in their homes, and even prepare for marriage. This expectation made a profound impression on my 18-year-old self. Between my experiences in California and Kathmandu, by the end of my gap year I had started to care deeply about female equality. This had become part of my self-identity. So at university I took on the role of women's officer at my college, representing undergraduate women's issues and rights. This turned out to be surprisingly controversial – at one point the women's notice board was burned down by people who apparently believed that feminism had no place there. My gap-year experiences had shaped my emerging sense of self. Who knows what kind of person I might have become had I travelled elsewhere or gone straight from school to university?

Our social self, the way other people view us, is central in adolescence. After my gap year, I was keen to advertise my emerging self to my new university peers, and they were doing the same as I was: hanging posters on their walls, listening to their preferred music, and joining university clubs – whether rugby, tennis, croquet or wine-tasting. They were expressing who they were. While self-identity continues to evolve throughout life, by the late teens most people have developed a sense of who they are and how they are – or would like to be – seen by other people.

The sense of self that stems from thinking about how we are seen by others is sometimes called the 'looking-glass self'. We imagine how we appear to other people and how they will judge us, and this might induce feelings of contentment, embarrassment, pride, shame or guilt. The process is complicated by the fact that we appear differently to different people – to family members who know us well compared with co-workers who know us less well, or know us for different qualities, for example.

Adolescents are more likely than younger children to compare themselves with others and to understand that others are making comparisons and judgements about them; they also begin to place higher value on these judgements. In other words, in adolescence the looking-glass self starts to play a larger role in the development of the self and we become increasingly aware of, and concerned about, the opinions of others.

One of my friends with teenage daughters told me that the most striking change he noticed in them around puberty was in their levels of embarrassment, especially embarrassment caused by their parents. Before puberty, if they were misbehaving in the supermarket all he had to do was to promise to sing their favourite song in reward and they would instantly behave. After puberty, the same promise became a threat: they could imagine nothing worse than their dad singing in public!

A brain-scanning study in 2013 led by Leah Somerville at Harvard University showed a striking effect of embarrassment on the body and the brain in adolescents. Each participant – child, adolescent or adult – lay in the brain scanner and was told that, at certain points during the scan, they would be observed by a peer (someone the same age as the participant) via a camera. These moments of observation would be indicated by a red light flashing on. Participants didn't have to do anything in the scanner – there was no task to perform, no stimuli to look at; instead, they were just told to think about the fact that their face was being observed by someone else when the red light flashed. In fact, though the participants didn't know this, there was no observer. But at the times when they *thought* they were being observed, or even just anticipated being observed, adolescents reported higher levels of embarrassment than either children or adults. In addition, skin conductance – the amount of sweat produced by the skin, which is a measure of stress and arousal – was heightened in adolescents who thought they were being observed, or anticipated being observed, more than in children or adults. Finally, thinking they were being watched was associated with greater activity in the adolescents' medial prefrontal cortex, a key region of the 'social brain' (the network of brain regions involved in understanding other people) that is involved in reflecting on the self. These physiological effects in the brain and elsewhere in the body suggest that thinking you are being observed, or anticipating being observed, by a peer makes you self-conscious, and these effects are more pronounced in adolescents than in other age groups.

Do you remember being a teenager? Perhaps you went through a phase of being acutely self-conscious, when you thought other people were constantly evaluating you, even talking about you. This is quite common in early adolescence (around age 11–14), when young people become increasingly aware that others have the capacity to evaluate them, and as a result may overestimate the extent to which this actually occurs. As they begin to question who they are and how they fit in with other people, young adolescents may become increasingly self-conscious, to the extent that they imagine an audience even if it doesn't exist. The term 'imaginary audience' was coined by the psychologist David Elkind in the 1960s. It describes the phenomenon whereby adolescents imagine that other people are constantly observing and evaluating them, even if this is not actually the case. Picture a 14-year-old girl not wanting to play board games with her parents and siblings because she knows her friends would think it is uncool – even though no one is watching her or would be likely to find out.

Many social media sites have been designed to appeal to young people's desire to share information about themselves, to be evaluated by others and to get a glimpse of other people's lives – Facebook, Snapchat and Instagram are good examples. However, more recent studies suggest that while our sense of the imaginary audience increases between childhood and adolescence, it remains quite high even in adulthood. Perhaps we are all, to some extent, overly concerned with what other people are thinking about us. Even though Facebook, for example, was set up by, and aimed at, young people, it has become increasingly popular with older people too. We all seem to be interested in how we are seen by others, and social media might be amplifying that interest and allowing us to express it more explicitly than in the past. Even after adolescence, our sense of who we are is constantly evolving throughout our adult lives.

Our sense of self does not depend solely on our looking-glass self – how others view us. It also stems in part from our own assessments of what we are like, formed on the basis of our own reactions to past and present events. My gap-year experiences in California and Kathmandu played a role in shaping my sense of self, with direct consequences for how I approached university. A related phenomenon is called 'introspection' or 'meta-cognition', which involves examining and having insight into our emotions and thought processes. What are you feeling right now? Perhaps you're feeling nervous or anxious? Or content or excited? These emotions might make you act and behave in certain ways; you might have insight into the emotions behind your behaviour – or you might not. The tendency to interrogate why we are feeling a certain way, and coming up with an account for our emotions, is a form of introspection.

Another aspect of introspection involves reflecting on how confident we are about our decisions and actions. When you give the answer to a question, you tend to believe that it's the right answer. But this isn't always the case: sometimes we know we are right; other times we make a decision or an action without really knowing why, or whether it was right or not. Imagine you've just completed a Sudoku puzzle or answered a difficult question in a general knowledge quiz – how confident are you that you got it right? People differ in the degree of insight they have into their behaviour and responses. Some people make decisions without much insight into why they took any particular decision or whether or not a certain response was correct – this is called low introspective ability. Other people are adept at knowing what they know. They know when a response they have given is correct or just a guess. These people have high introspective ability.*

How does the ability to introspect develop during adolescence? Leonora Weil investigated this question in a study carried out in my lab in 2012. Participants aged between 11 and 41 years were asked to carry out a fiendishly difficult perceptual task, which involved looking at two pictures on a computer screen, one after the other. Each picture contained six identical shapes, and in one picture one of the six shapes was very slightly brighter than the rest. Participants were asked to identify which of the two pictures contained the brighter shape, and were then asked to rate how confident they were about their decision. The task was made harder by the fact that each of the pictures was shown for only a fraction of a second – participants reported it felt like they were guessing, even though most of the time their answers were correct. It's a bit like when you're at the optician and being asked to read letters that are barely visible to you: often it feels like you're guessing.

* How confident you are about your performance on cognitive tasks or decisions is not necessarily related to the accuracy of your performance or decisions. Some people with high confidence are falsely confident in their own judgements.

In our experiment, we found that, although performance on the perceptual task was similar at all ages, late adolescents and adults were better at knowing whether or not they had chosen the correct picture than younger adolescents were. In other words, people's ability to identify their own accuracy levels – their *introspective ability* – increased with age during adolescence and levelled out in adulthood.

As well as enabling you to reflect on how others see you, and on your own responses to questions and challenges, your sense of self involves the ability to think about what you might do in a given situation – about alternative possibilities for how you might act. Perhaps you don't hear your alarm in the morning, so you oversleep. You're going to miss the train, which means you'll be late for an important meeting. What do you do? Perhaps you make a phone call to warn your colleagues you might miss the meeting, and ask someone else to go in your place. You might look at the bus timetable, or call for a taxi, or ask a friend for a lift. However you respond, your intentions – in this case, to get to the meeting – are driving you to make certain decisions and actions.

With Suparna Choudhury, who was a PhD student at the time, and Hanneke den Ouden and other collaborators, I investigated the development of the ability to think about what we might do in a given situation. We called this 'intentional causality' because we were interested in how our intentions cause our actions. We asked volunteers to think about what action they would take, given a particular intention (for example: 'You are at the cinema and have trouble seeing the screen; do you move to another seat?'). As our particular interest was in development during adolescence, we included in our study group both adolescents aged 12–18 years and adults aged 22–37 years. Participants' brains were scanned while they answered the series of questions about what action they might take given a particular intention. In a comparison task, participants were asked about what might happen given some physical event (for example: 'A huge tree suddenly comes crashing down in a forest; does it make a loud noise?'). These tasks are similar – they involve causes and consequences – but the first involves your own intentions, while the second involves physical occurrences which you don't cause.

We found interesting differences between the two groups, adolescents and adults, in activity within the 'social brain' network. This network is involved in understanding other people's intentions and emotions, and it includes brain regions involved in thinking about the self. When answering questions about their own intentions, as opposed to questions about physical events, the adolescents in our study used one part of this brain network – the medial prefrontal cortex – more than the adults did. Remember, this was the region activated in Leah Somerville's study in which participants thought they were being observed in the scanner by a peer, and became self-conscious. In contrast, the adults used another part of the social brain – the temporo-parietal junction – more than the adolescents did. The study demonstrated that between adolescence and adulthood there is a change in the pattern of brain activity associated with self-reflection.

It's difficult to establish *why* this happens, but one explanation is that adolescents and adults use different cognitive strategies to complete these tasks. In other words, the mental approach to thinking about the self might change with age, and this would be reflected in different patterns of brain activity. Adolescents, it seems, might think about the self more consciously when contemplating themselves. Perhaps for adults, thinking about the self has become more automatic, and involves less effortful, conscious thought; they have greater reservoirs of stored memories and experiences, and can dip into these resources when deciding how to respond in social situations, rather than consciously thinking about themselves. Perhaps the adult brain allows us to access our memories and experiences to make assessments about ourselves in a way that the adolescent brain doesn't yet do to the same extent.

This is an explanation that has often been used to account for the differences in brain activity between adolescents and adults, as we shall see in later chapters of this book. However, it comes with a warning: it is not really valid to make inferences about psychological processes on the basis of activity patterns in the brain. This is because there isn't a one-to-one mapping between brain regions and specific psychological processes – each brain region is involved in multiple processes. To give an example, the temporo-parietal junction is involved in paying attention as well as thinking about other people's minds. Making a 'reverse inference' from the brain region activated to the psychological process is therefore potentially misleading. We cannot know what people are experiencing on a psychological or perceptual level from the parts of their brains that are activated. But we can speculate, and make predictions that can be tested in future experiments.

The sense of self can be altered by damage to single brain regions. However, damage to any single brain region does not, as far as we know, completely obliterate the sense of self. This makes sense: it's unlikely that something as complex as the sense of self resides in a single brain region. Rather, there are many different aspects of the self – including the ability to distinguish self and other, the looking-glass self, the ability to introspect, and our cumulative store of memories and experiences. It's likely that each aspect emerges from more than one different neural system. These different neural systems interact with each other to produce the complex set of behaviours, perceptions, dispositions and character traits that make up the (whole) self. For many of us, a deep and complex sense of self, particularly of our social self, has its origins in adolescence. And in developing that social self, one group of people stands out as being exceptionally significant: our friends; other adolescents – people like us.

3

Fitting in

When you think back to your teenage years and early twenties, what comes to mind? Perhaps you have vivid memories of school, holidays, discovering and learning about things for the first time, exams, parties and romance. For many people, most memories of these years focus on our friends and our social worlds. There might be good memories – forming strong friendships, celebrations, get-togethers, connecting with people; there might be some bad memories – being excluded from a social group, being teased or bullied, splitting up with a girlfriend or boyfriend. Whether positive or negative, memories of adolescence often involve our peers.

In adolescence, friends matter. Studies going back many decades have found, and new studies continue to find, that friends are more important to us during adolescence than at any other stage of life. And it is particularly important to adolescents to be accepted by their peer group. This has many consequences, including an especially strong susceptibility to peer influence; and this in turn has an important impact on adolescent risk-taking and decision-making.

The stereotypical image of adolescence includes a taste for engaging in risky and novelty-seeking behaviours: smoking cigarettes, taking drugs, drinking alcohol, dangerous driving. Not all adolescents are risk-takers or novelty-seekers, of course, but there is evidence that risky behaviours peak in adolescence. Beneath the stereotype, though, lies a more complicated picture. To understand risk-taking in adolescence, it's important to think about the circumstances in which adolescents typically take risks. Usually it's not when they're on their own; most adolescents who take these kinds of risks do so when they're with their friends.

The effect of peers on risk-taking was shown very clearly in a series of experiments in the mid-2000s carried out by Laurence Steinberg and colleagues. They designed a driving video-game that participants played in the lab. In this computer game, called the Stoplight Task, each participant drives a car around a track, and the goal is to get to the end of the course in as short a time as possible. Every so often there are traffic lights and, if the light happens to be yellow, the participant has to make a decision. They can wait until the light goes green again, which will cost them time; or they can take a risk and drive through the yellow light. Sometimes the risk pays off: the light stays yellow and the participant gets round the track more quickly, earning more points in the process. But at other times the light turns red; this leads to a crash, and the participant loses time and points.

Participants play the game twice: once alone and once with friends of the same age. In this study, 306 participants, comprising adolescents (13–16 years), young adults (17–24 years) and adults (25 years and over), carried out the Stoplight Task.

When they were on their own, each age group took around the same number of driving risks. In the second version of the game, the researchers asked each participant to bring a couple of friends with them to the lab, and the friends stood behind the participant when he or she was playing the driving game. The friends were told that they could call out advice about whether to allow the car to keep moving or to stop it. The player was instructed that he or she could choose whether or not to follow the advice of his or her peers.

So, the same participants played exactly the same driving game as before – but there were stark differences between the two rounds.

When in the presence of their friends, adolescents took almost three times as many risks as when they were alone, and young adults took nearly twice as many risks. In adults, the presence of peers had no impact on risk-taking.

There are two important findings from this study. First, it shows that adolescents, young adults and adults all take around the same number of risks when they are alone, in optimal conditions with no distractions. It's important to note this, because it means that adolescents don't always take risks, contrary to the stereotype. Second, it shows that a critical factor in risk-taking for adolescents, and to a certain extent for young adults, seems to be the presence of peers, whereas this is not the case for adults aged 25 and over.

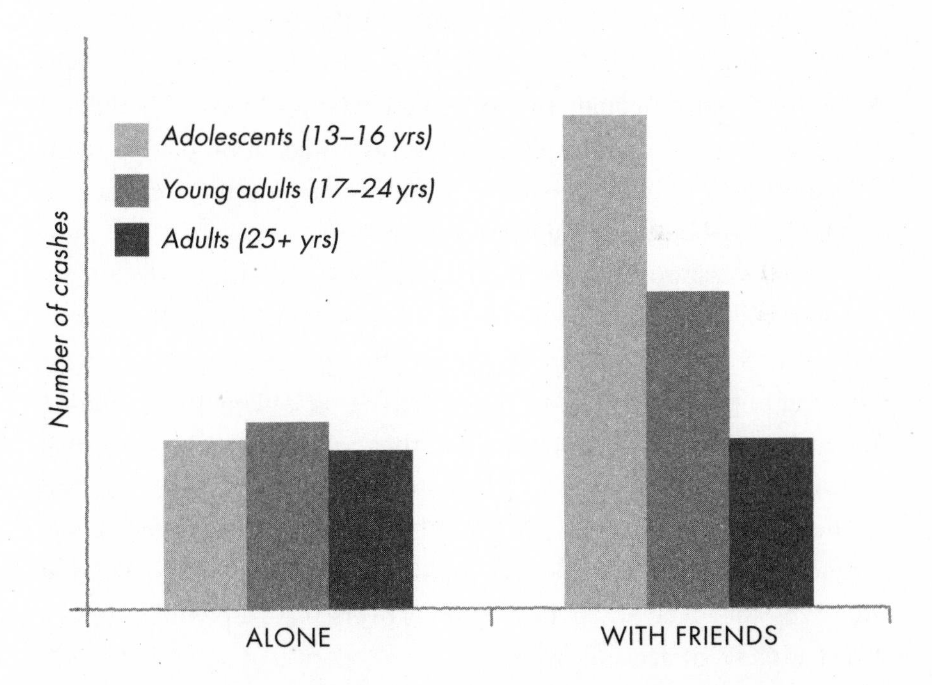

Risks taken in the Stoplight Task. Adolescents, young adults and adults all take around the same number of driving risks when alone, but adolescents and young adults take many more risks when being observed by friends

This lab-based finding is confirmed by real-life observations, for example in data from car insurance companies. If you have ever had a car accident and made a claim on your insurance company, you will know that you have to tell them the precise circumstances in which you had the accident. What you might not know is that the insurance company will often enter these details – anonymously – into a large database. The accumulated data can then be analysed to look at the circumstances in which crashes are most likely to happen.

When scientists have analysed car insurance data, they have found two key findings relating to young drivers. First, people aged 16–25 have more car accidents than people aged 26 and over. That's why young people's car insurance premiums are higher. These days, insurers offer special schemes in which 'black boxes' are fitted in cars and record young people's speed, braking and accelerating, so that their premiums can be reduced if they drive safely. Second, the datasets show that young people are more likely to crash when they have a passenger in the car with them. In contrast, this is a protective factor for adults 26 and over, who are less likely to crash if they have a passenger than when they're alone in the car.

What is it about the presence of peers that makes adolescent drivers more likely to crash their cars? One obvious possibility is that adolescents find passengers more distracting than adults do. Perhaps chatting to passengers makes concentrating on driving more difficult for adolescents. Another possibility is that risk-taking behaviour such as fast driving might be seen as being cool and socially desirable, and that adolescents are responding to this social pressure. Many people, including Kate Mills, Anne-Lise Goddings and I, have suggested that this need for social acceptance by one's peers plays a pivotal role in a lot of adolescent decision-making.

There's a long history of social psychology research on the importance of peers in adolescence. In one series of studies published in the 1990s, US researchers provided children aged 9–15 years with a pager.

At random times of the day, every day for a week, the children were paged and asked questions about where they were, what they were doing and who they were with, and about their current mood and feelings. The same children were tested again, a couple of years later. The researchers analysed the huge amount of data from the paged responses, and looked at how the responses changed with age. These studies mapped out the trend among American adolescents (especially girls) of spending more and more time with their friends, and less time with their parents and other family.

There were gender differences, too. Adolescent girls reported spending more time with peers and less time with parents compared with childhood, while the amount of time they spent alone didn't change much. Adolescent boys reported spending more time alone, the same amount of time with friends and less time with their parents. The difference is interesting; but it's important to remember that these are averages, and not true for all girls and all boys. In addition, this survey was first carried out in 1991; if it were replicated today, the results would most likely be different. In 1991, there were no social media, fewer interactive video-games and far fewer mobile phones. These days, adolescents spend a lot of time interacting with their friends over the internet and by instant messaging. How would this be classified? It's certainly social interaction, but it's different from face-to-face interaction.

More recent surveys have found marked cultural differences, with adolescents in some cultures, for example in Japan, South Korea and India, spending just as much time with their families as adolescents as they did when children. This emphasizes the importance of taking culture into account when thinking about adolescent development. Different cultures exert different pressures on adolescents. It might be permissible for adolescents to behave a certain way in one culture but not in another.

∞

It's not just that adolescents in many Western cultures choose to spend more time with their friends than they did when they were younger (and of course they have more independence than when they were younger to do so). What is particularly significant is the weight of peer pressure at this time; the fact that, for adolescents, the opinions of their peers become more important to them than those of family members.

Insight as to why these opinions become more important can be found in a study from the late 1980s. When interviewed about friendships, children aged 10–13 years reported that friends provided companionship, stimulation and support, but they did not feel that peer acceptance influenced their evaluations of themselves – their self-worth. In contrast, adolescents aged 13–17 years reported that evaluations by their friends affected their feelings of social or personal worth, and that being rejected by peers indicated their unworthiness as individuals.

As young people move from childhood into adolescence they start to care more deeply about what their friends think about them, and this affects their sense of self-worth. What drives this increased sensitivity to friends' opinions and judgements? It is important for adolescents to become independent from their parents, if they are to thrive later in life; but it is also important to integrate into their peer group, for reasons we'll consider later in this chapter. In our theory, Kate Mills, Anne-Lise Goddings and I have suggested that adolescents feel a particular concern about being socially excluded by their peers and a strong desire to be included in their social group. This might result in adolescents being especially sensitive to what their friends think of them, so that they can integrate this information into their views of themselves and adapt to their friends' expectations and social norms, which would increase their chances of being accepted into the social group.

A few years ago, Catherine Sebastian and I carried out a study to look at whether adolescents are particularly negatively affected by being socially excluded. We conducted a lab-based experiment in which social exclusion was simulated using a computer game called Cyberball. This was originally devised over twenty years ago by Kip Williams at Purdue University in the United States. It's an online ball-throwing game in which the participant plays with two other players who are not present but are represented by cartoon icons on the screen. As soon as the game starts, the participant is thrown the virtual ball by one of the other two 'players'. The participant can then choose which of the two other players to return the ball to, by clicking on that player with the computer mouse. The player with the ball either returns the ball to the participant or throws it to the other player. This goes on for several minutes.

The game is programmed so that in some trials – called 'inclusion trials' – the participant is included in the ball game and is passed the ball around a third of the time. In 'exclusion trials', the other two players throw the ball back to the participant at first, but then stop doing so. For the rest of these exclusion trials, the other two players throw the ball to each other, but exclude the participant from their game of catch.

This game has been used in many different experiments with adults, which have shown that even in this simple and short online game, mood is lowered and anxiety is increased after exclusion trials. We don't like to be excluded, even if it's in an online ball game with players we will never meet. Catherine Sebastian and I wondered what would happen if we asked adolescents to play Cyberball. We hypothesized that their mood and anxiety levels would be even more affected by social exclusion than those of adults.

We studied groups of young adolescents (aged 11–13), mid-adolescents (aged 14–16) and adults (aged 22–47). The participants came to the lab in London and we told them they were playing the ball game with real people their own age, online, but this was just a cover story. In fact, we had programmed the game ourselves and the other players weren't real, so that we could control precisely what happened during the course of a game. We checked – by asking them afterwards – that all our volunteers did indeed believe they were playing with real people online, because this was important for our study.

In the adult group, mood was lowered and anxiety was increased after being excluded by other players, just as Kip Williams and his colleagues had found previously. Both adolescent groups showed the same pattern as adults. However, both young and mid-adolescents reported significantly lower overall mood than adults after social exclusion. Young adolescents also reported higher anxiety than adults after exclusion. In other words, it seems to be the case that adolescents are *hypersensitive* to social exclusion.

This study provides some evidence that adolescents are especially sensitive to how they fit into their social environment – in this case, being included by other people in a game. The same heightened sensitivity has been observed in adolescent rodents. If rats are exposed to stress, including social stress (such as being isolated), in the twenty or thirty days of 'adolescence' between starting puberty and becoming an adult, some of the negative effects are longer-lasting than, and qualitatively different from, the effects of exposure to stress at other periods of life.

For example, exposing rats to social isolation – in which they are housed alone and have no contact with other rats – during adolescence increases the likelihood of depressive behaviours and is associated with changes in the structure of the prefrontal cortex in adulthood. Social isolation affects the way the brain develops in adolescent rats more than it does in adult rats. Male* adolescent rats exposed to social instability – in which they are isolated for an hour

* Most rodent studies are carried out with male animals. This is because fluctuations in hormones make it more challenging to study female animals. As a consequence, little is known about brain and behaviour in female rodents.

each day and experience a change of cage partner – also show changes in behaviour and hormone production. Compared with rats that had stable social environments during adolescence, they are more anxious and less socially interactive when they reach adulthood, and exhibit abnormal sexual behaviours. The rats exposed to social instability also have lower testosterone concentrations as adults.

These studies suggest that for (male) rats, social isolation and social stress during adolescence have far-reaching consequences in respect of brain structure, hormone levels and behaviour in adulthood. Do we see any analogous effects in human studies? This is a tricky question to study, because we can't deliberately put human adolescents in truly socially stressful environments and measure their reactions – quite rightly, no university ethics committee would allow it. Cyberball is one thing; actually isolating an adolescent for a long period of time would be quite another. So instead, researchers observe humans who, sadly, are already exposed to severe levels of social stress in their lives.

These studies have shown that, just like adolescent rats, human adolescents who experience social stress suffer behavioural consequences in adulthood. Adolescents in very socially unstable environments – moving between foster homes and children's homes, for example, and possibly experiencing violence and chaos in the home – tend to be in poorer physical and mental health than adolescents whose social worlds are relatively consistent and stable. The consequences of social instability in adolescence – in both rats and humans – can be so detrimental that mechanisms and behaviours promoting peer acceptance can be considered adaptive. That is, it might be evolutionarily beneficial for adolescents to do their utmost to be accepted by their peer group, so as to avoid being socially isolated.

This brings me to the idea that the fear of being socially excluded influences many decisions that adolescents make. One of the questions I'm often asked by parents and teachers is why sensible, well-educated young people who know about the health risks of – say – cigarettes nevertheless take up smoking. Some people put this down to not really understanding risk, or to feeling invincible, but there's not much evidence for either of these. When asked in a laboratory setting to estimate the likely negative outcomes of risky behaviours, adolescents sometimes even tended to overestimate the risks. In addition, there's not a lot of evidence that adolescents feel invincible and unaffected by risk.

To understand why adolescents take risks, we have to look at the contexts in which risk-taking occurs. Adolescents are more likely than children and adults to make risky decisions in so-called 'hot' contexts, such as when they're with their friends or when emotions are running high.

∞

One summer recently, I was driving home from dinner at my parents' house in Oxford. The drive between Oxford and my house is along quiet countryside roads. It was about 1 a.m. and I noticed a young woman walking along the side of the road. This road had no pavement, it was dark, and the next village was miles away. I stopped and asked her if she'd like a lift somewhere. She took me up on the offer and, on the drive to her parents' house, told me she'd been at a party at a pub in a town a few miles back. She could have left the party earlier, in time to get the last bus home, but she was having so much fun with her friends she decided to stay – a classic 'hot' context! She had no money for a taxi, there were no night buses, and her phone had run out of battery so she couldn't call anyone. So she decided to walk 8 miles home in the middle of the night.

In the cold light of day it's easy to pass judgement, but I made similar decisions when I was that age (my passenger told me she was in her late teens). When we were teenagers, on holiday in France, my friends and I regularly hitch-hiked all over the place, day and night, without giving a second thought to its potential dangers. I wouldn't dream of doing it now.

But maybe I've become too risk-averse as an adult. Taking risks isn't always a bad thing: it can lead to new experiences, learning and personal development, and it can be fun. Some risky decision-making is necessary in development and indeed throughout life. Risk-taking can be useful in academic contexts. Putting your hand up in class, guessing answers in a test, public speaking, debating – these all entail risks that can result in new learning and gains in confidence. Some risks, such as staying out late at a pub with friends and missing the last bus home, as my passenger did, might result in peer acceptance. My passenger got to spend longer with her friends – and she got home safely: the risk paid off.

Kate Mills, Anne-Lise Goddings and I proposed a 'see-saw' model to show how social factors can affect risky decision-making. Every time we make a decision we weigh up potential good and bad outcomes. For example, when deciding whether to use a phone when driving, there are clear reasons not to: it's against the law and could cause an accident. However, a number of social factors make us more or less likely to use our phones despite knowing the risk of negative consequences. What would other people in the car (or on the street)

Should I use my phone when driving?

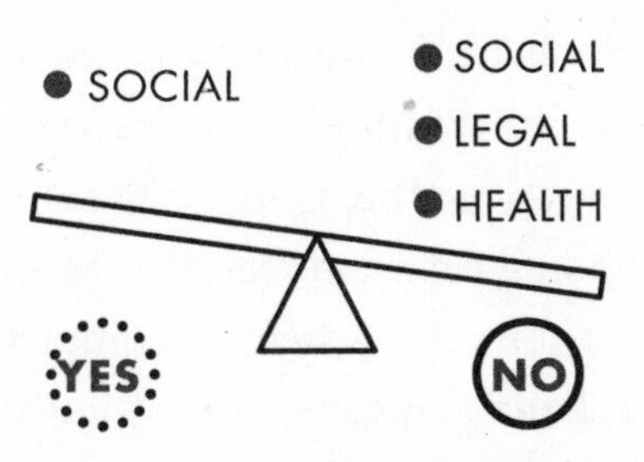

The see-saw of decision-making. It's clearly a really bad idea to use your phone while driving. Yet some people decide to do so, because of the social pressure to respond to messages.

think about us using our phones? But what if we really want to communicate with someone? We might need to let someone know we are running late, or share some exciting news. These social factors affect decision-making more for adolescents than other age groups.

Every day we make many decisions, mostly minor, occasionally major, some involving risks, others not. Whenever you make a decision, you weigh up its pros and cons. Another example, like deciding whether or not to use your phone while driving, might be speeding. Imagine you're at work and want to get home for a family dinner, but it's getting late. Do you drive over the speed limit to get home in time? Several factors feed into the decision-making process: the potential advantages of speeding, such as the satisfaction of getting home when you planned to, and the feeling of pleasure you might get from the experience (some people like to drive fast), as well as the potential disadvantages, such as the possibility of having an accident (health) or being caught by a speed camera or the police (law).

In addition to these pros and cons, there's also another potential element in the decision process – the social factor. Your family will be pleased to see you at dinner, and annoyed if you're late. But what will your friends think if you get caught speeding? That could be embarrassing at work, too. What if you have a couple of friends in the car with you, encouraging you to put your foot down? Maybe you want to impress them?

Of course, it's not just adolescents who are influenced by others – we all are. We behave differently when we're in a group compared with when we're on our own. The field of group behaviour in social psychology came about because researchers wondered why otherwise well-behaved adults sometimes behave completely differently when they're in crowds at football matches – the *football hooligan* stereotype – and many decades of research have now gone into investigating the phenomenon. It's not just football fans; most of us behave differently in crowds, especially when emotions are involved, perhaps because we feel less personally responsible for our behaviour when many people are doing the same thing.

People also perform differently on cognitive tasks when they're being watched by someone else compared with when they're carrying out the same task unobserved. This is called the *audience effect*. An audience makes hard tasks even harder, and people perform worse at these when being watched than when alone. Picture reverse-parking your car while someone in another car is waiting for you to move out of the way – often, people in this situation feel under a lot of pressure, even though the actual physical task is no different from when you're doing it on your own. In contrast, people do better at easy or well-rehearsed tasks when being observed than when performing the same tasks alone. Both cases may have to do with the increased stress introduced by an audience: a small amount of stress improves performance in an easy task, but the combined stress of carrying out a challenging task and being observed impedes performance.

We are all influenced by other people. What I would argue is that adolescents are *especially* susceptible to social influence. In terms of the decision see-saw, this means that social factors weigh in particularly heavily for adolescents; adolescent decision-making is more driven by the need for peer acceptance and the desire to avoid being socially excluded. This means that in some situations adolescents might take risks when they're with their friends that they wouldn't take when on their own.

∞

Let's take the example of experimenting with drugs. Imagine a 15-year-old girl, doing well at school, close to her family and popular with her class-mates. One Saturday evening, she meets a group of her friends in town. They have all taken Ecstasy and offer her some. She knows that Ecstasy can be dangerous. She is also worried about being caught and getting into trouble. But despite all these things, she accepts the pill.

Even though adolescents understand the health risks of experimenting with drugs, there is also the *social risk* to consider: saying no to a drug when all your friends are taking it, and potentially being ostracized by your social group as a consequence, might be perceived as more risky than accepting the tablet. Adolescents might be particularly susceptible to this kind of social pressure because it's acutely important to them not to be rejected by their peer group.

Peer influence doesn't always lead to risk-taking. In fact, in some situations the desire to avoid peer rejection might lead adolescents to *avoid* taking a risk. Teachers I speak to often bemoan the fact that intelligent and knowledgeable students won't take risks in the classroom. The same adolescent who accepts a drug from her friends might not raise her hand to answer a question in class for fear of looking stupid (or too clever!) in front of her friends. In this way, social pressure can lead to risk aversion as well as risk-taking.

The important, if obvious, caveat to all of this is that there are large individual differences in adolescent behaviour. Some adolescents don't take many risks at all and don't seem to worry about what their peers think of them. Little is known about what causes these individual differences, or what the consequences are. Is it beneficial to be influenced by peers in order to fit in with the group? Does this make you a more successful adult – socially, professionally?

There are some interesting studies suggesting that risk-taking in adolescence might be associated with certain forms of success in adulthood. One study that began in 1965 involved assessing 1,000 Swedish children from the age of 10 for more than three decades. In one analysis, early rule-breaking behaviour in adolescence, such as staying out late without permission, cheating in an exam, getting drunk or shoplifting, was associated with entrepreneurial careers in adult men (but not in women). So perhaps there are adaptive advantages to being a risk-taking adolescent.

A few years ago, Lisa Knoll, a post-doctoral researcher in my lab, and I designed an experiment to look at social influence on risk perception. We got permission to run it in the Science Museum in London, where we set up three laptops in a room at the back of one of the galleries. Lisa and a team including several students and interns then decamped to the museum, where they carried out the experiment for just over three weeks. In that time, they managed to test more than 660 visitors aged between 8 and 59 years on the computerized task we had constructed. An impressive number!

The task involves being shown a series of risky scenarios, such as crossing a street on a red light, riding a bike without a helmet or walking down a dark alley. Participants were asked to rate the level of risk they attached to each scenario. The scenarios were chosen because they carry a small to moderate amount of risk – and, critically, the assessment of that risk is subjective: different people perceive the same situation as carrying different amounts of risk. This was crucial, because what we were interested in was whether people's perceptions of risk would change when they were told how risky other people thought the situation was: that is, we were interested in the *social influence* effect.

Having registered their own risk ratings, participants were shown the risk ratings given by other people for the same scenarios. They were told that these were ratings given by other people – either adults or teenagers – who had taken part in the experiment, but in fact, this was just a cover story: the risk ratings provided were randomly generated by the computer. Participants were then asked to rate the riskiness of each scenario for a second time. We were interested to see how much participants changed their risk ratings after they had seen the ratings provided by other people.

There were three main findings from this study. First, the perception of risk reflected in the first ratings differed between age groups. Children aged 8–11 rated the situations as more risky than other age groups. Interestingly, there was no evidence that young adolescents (aged 12–14) or mid-adolescents (aged 15–18) perceived the situations as less risky than adults did. This supports the idea that adolescents understand risk and suggests they do not see themselves as immune to it.

The second finding was that all age groups showed a social influence effect: everyone shifted their risk perception in the direction of other people's risk ratings. This social influence effect was highest for children, and lowest for adults aged 26 and over. But even these older adults changed their risk ratings to be more in line with other people's ratings. This supports the idea that we are all influenced by other people.

The third finding was the most interesting to us. In our experiment, participants were told that the (actually fictitious) risk ratings provided by other people were either from all the adults who had taken part in the study, or from all the teenagers who had taken part. We could therefore look at whether participants were more influenced by the opinions of adults or those of teenagers. As the chart opposite shows, whereas children (8–11) and adults (19–59) were more influenced by *adults'* opinions about risk, young adolescents (12–14) were more influenced by the opinions of *teenagers*. Mid-adolescents (15–18) showed a similar level of influence by adults' and teenagers' opinions about risks (the small difference shown in the chart was not statistically significant), suggesting that this is a transitional stage in development.

We recently replicated these effects in a completely new study involving a different set of 590 participants aged between 8 and 59. In science it is vitally important to know that an effect is replicable. The results of both studies support the notion that early adolescence is a pivotal phase in which individuals begin to question the authority and experience of adults and place higher value on the opinions of other teenagers than on those of adults. They also suggest that health advertising aimed at young people, rather than focusing on the risks of

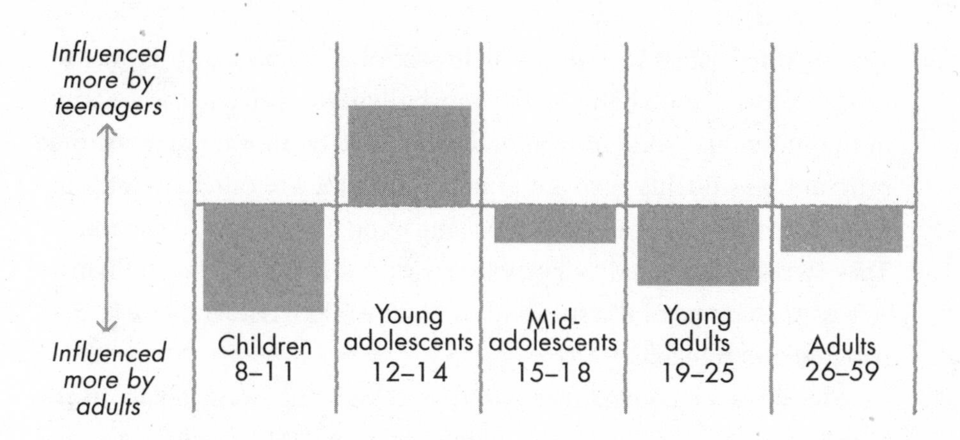

The social influence effect in risk perception

dangerous behaviours such as smoking, binge-drinking and experimenting with drugs, should perhaps focus on social norms and peer influence. What really seems to matter at this age is what friends and contemporaries think.

A 2016 study, carried out by researchers at Yale and Princeton universities, demonstrated the real-world implications of this peer influence effect. This study looked at the influence of social norms on bullying behaviour and conflict in schools. It covered fifty-six middle schools (with pupils aged 11–16 years) in the state of New Jersey, half of which were assigned at random to an anti-bullying programme. In this programme, groups of between twenty and thirty-two students in each year group were invited to participate in an anti-conflict workshop, which involved a trained researcher working with the students to help them understand the negative effects of bullying.

The students on the programme were encouraged to lead grassroots anti-bullying campaigns in their schools and become the public face of opposition to bullying. In one activity, the programme students were encouraged individually to design anti-bullying poster campaigns and slogans. Each student's photo and name were included next to the slogan he or she had created, and the posters bearing the slogan were displayed around their schools. The aim of this was to create an association between the anti-bullying slogan and the identity of the individual student who had created it. In another activity, the programme students gave out orange wristbands to other students in their schools who were observed engaging in friendly behaviours. These wristbands acted as a visible reward for action against bullying. Across the course of the study, more than 2,500 wristbands were distributed and tracked.

Measures of conflict behaviour and bullying were taken from all fifty-six schools across the school year, and the results were remarkable. Compared with the control schools, in which no special anti-bullying programmes had been introduced, reports of student conflict in the schools that had implemented the student-led anti-bullying programme had fallen by 30 per cent. In each case the intervention, based on a small number of students publicly opposing bullying and conflict, had successfully spread new anti-bullying attitudes through the school. Furthermore, the researchers measured social connections in each school, asking all students to report which students in the programme they had chosen to spend time with in the previous few weeks, thereby generating an indication of the popularity of each student in the programme. This analysis identified a number of highly connected students, and the results showed that the effect of the anti-bullying programme was stronger when more of these highly connected students were involved in the campaign. In other words, when the anti-bullying campaign was led by the more popular students, it had a greater positive effect on behaviour. It seems that popular students have a greater influence than others on social norms and behaviour in schools. The study reveals the real-life power of peer influence in changing social norms of acceptable behaviour and conflict in schools.

∞

Young adults emerging from adolescence need to be equipped to navigate the complexities of their social world. Anthropologists have accordingly suggested that adolescence is a time of particular cultural susceptibility, meaning that adolescents might be more likely than other age groups to pick up on cultural norms and behave according to local cultural rules and expectations. Suparna Choudhury, who did her PhD in my lab and subsequently changed fields from neuroscience to anthropology, has argued that the adolescent brain might change in response to the establishment of 'cultural niches' during adolescence.

Imagine a child who moves from one country to another – or even from one school to another – during their teenage years. It's socially important for them to fit in with their new peers and, to achieve this, they need to pick up on the cultural norms for their age group and to try to fit in with their new friends. A much younger child, on the other hand, has their parents to help them assimilate and is more likely to do what their parents do. Adults are more likely to have their original culture ingrained in them, and so to be slower to adapt to a new culture.

While the need for social acceptance plays a key role in adolescent risk-taking behaviour, it's not the only factor. The adolescent social environment is different from that of children and adults in various ways. In many school systems, the transition from primary to secondary school occurs around the onset of puberty, placing children in a new environment, often with a different and larger set of peers and different structures of learning. At this age, children move from being the oldest in the school to the youngest. Adolescents are also exposed to more novel situations – spending more time by themselves or with friends, getting to school by themselves, deciding who to hang out with, which classes to take, and so on – than they were likely to encounter as children. For perhaps the first time, they are making many more of their own decisions, independently of their parents or teachers. There is more opportunity to experiment and explore, and this new-found freedom enables adolescents to take more risks than before.

Putting this all together, it's clear that there are many factors, environmental as well as biological, to be taken into account in explaining 'typical' adolescent behaviour. It is also increasingly apparent that what is going on inside the adolescent brain is both complex and markedly different from what, until recently, was assumed to be the case. It's time to look more closely at what is happening inside the brain during these years of trying out, moving up and fitting in.

4

Inside the skull

Sometimes my sons astonish me. Last year, in 2017, they turned 12 and 10, heading rapidly towards the fuzzy boundary between childhood and adolescence. It feels as if, in the blink of an eye, two tiny babies have become thinking, feeling, complicated people, with their own individual personalities and motivations and idiosyncrasies, their own wit and ripostes. I don't remember it happening: it has been such a gradual, intangible process. And the next ten years will see many more profound changes.

The transition in humans from birth to adulthood *is* astonishing. How does a newborn baby, who enters the world with so little understanding, competence or independence, eventually develop into an

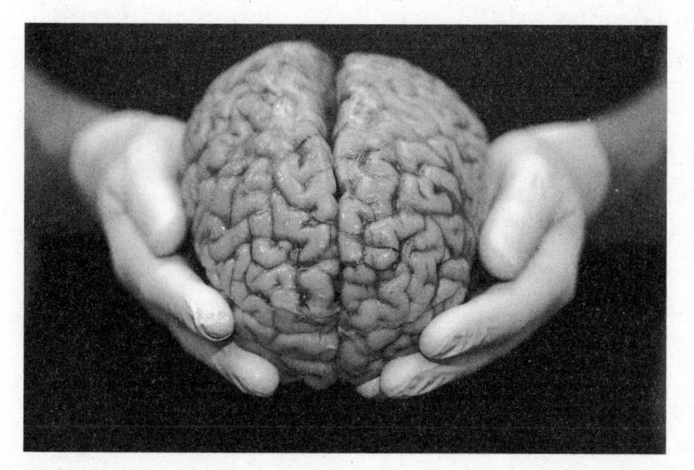

The human brain

autonomous, capable adult, no longer reliant on caregivers for its survival? This transition seems impossibly enormous, both physically and mentally. For centuries, scientists have tried to understand how the human mind – and the human brain – develops. At this point I need to tell you how this extraordinary organ is constructed, and the best way to do this is to relive for you that initial experience of holding a human brain in my hands.

Following the instructions of the lecturer, I used a scalpel to dissect the brain and expose its inner structure. The folds that can be seen as lines on the surface plunge deep inside and undulate across the cortex, the outer layer of the brain. The cortex is made up of four lobes, which are separated by deep crevices called *sulci*. According to the traditional textbooks, each lobe is responsible for a different type of behaviour or psychological process. In reality, this is too simplistic. Every lobe contains many different regions, each of which is involved in multiple different behaviours. For example, the temporal lobes – on either side of the brain, extending from your temples up and past your ears – contain regions including the auditory cortex, which processes sound information, areas involved in the understanding of language, and the superior temporal sulcus, which processes faces and other forms of social information.

Each region is connected to every other region in an incredibly complex network, and different parts of the brain are in constant communication with each other. There is a network of freeways in Los Angeles, comprising dozens of huge roads crossing one another and interconnecting. Whenever I have driven through it I've thought about the brain's network of connections – although the brain is infinitely more complicated.

Cutting further, deep into the centre of the brain, I discovered the subcortical regions. I'd read about these in textbooks and scientific papers: the amygdala, which processes emotion; the hippocampus, which stores memories; the basal ganglia, which are a set of small

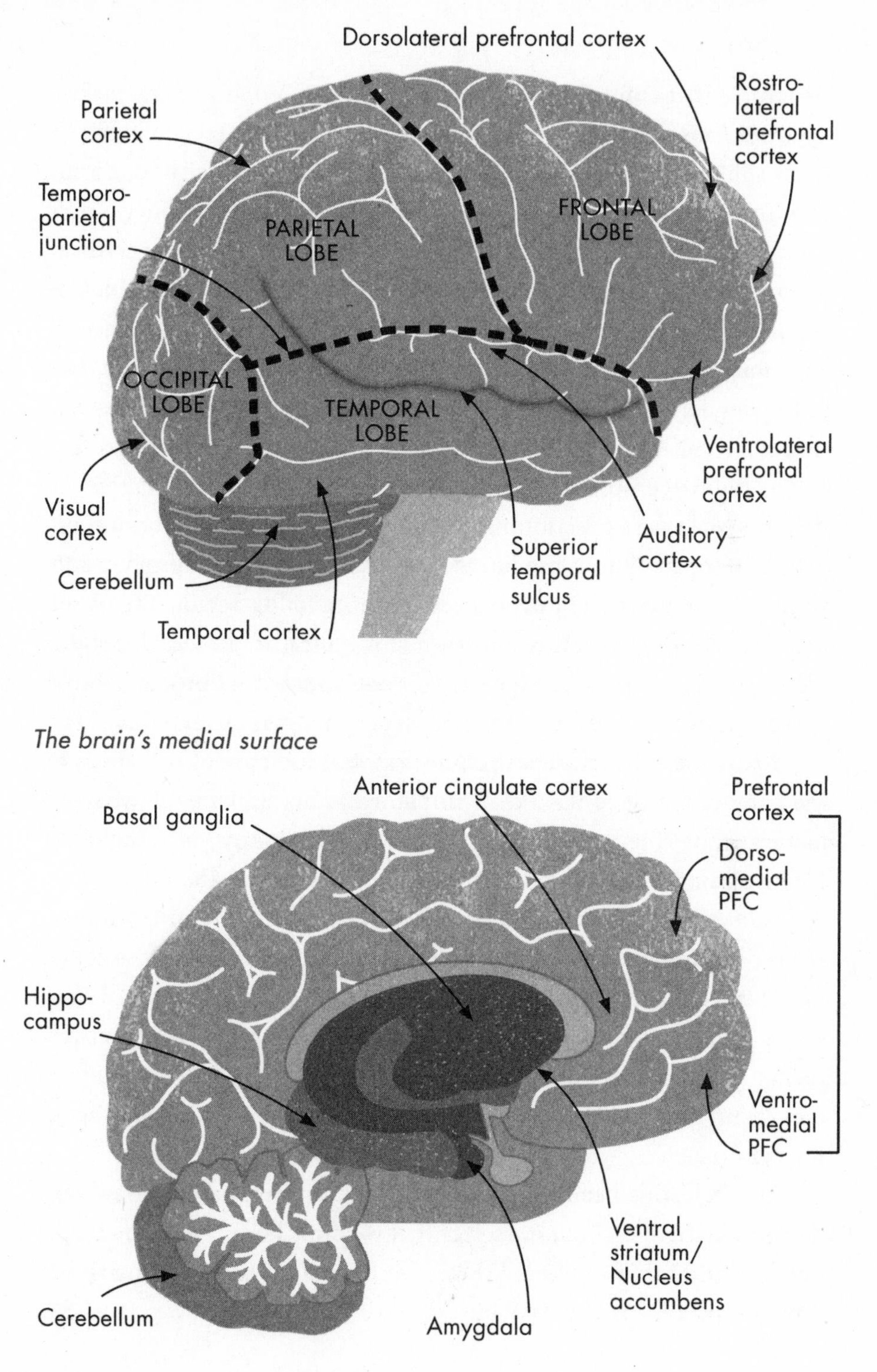
The brain's lateral surface
Dorsolateral prefrontal cortex
Rostro-
lateral
prefrontal
cortex
Parietal
cortex
Temporo-
parietal
junction
PARIETAL
LOBE
FRONTAL
LOBE
OCCIPITAL
LOBE
TEMPORAL
LOBE
Ventrolateral
prefrontal
cortex
Visual
cortex
Cerebellum
Superior
temporal
sulcus
Auditory
cortex
Temporal cortex
The brain's medial surface
Anterior cingulate cortex
Prefrontal
cortex
Basal ganglia
Dorso-
medial
PFC
Hippo-
campus
Ventro-
medial
PFC
Ventral
striatum/
Nucleus
accumbens
Cerebellum
Amygdala

structures that control movement and learning. What I hadn't picked up from those textbooks and papers is the visual difference between these subcortical regions and the cortex. The cortex is fairly uniform, vast and folded. The subcortical regions, however, are distinct – tiny discrete structures – and slightly darker than the cortex. The amygdala, from the Greek word *amygdale*, meaning almond, really does look like an almond; the hippocampus, from the Greek *hippos*, meaning horse, and *kampos*, meaning sea monster, has the shape of a tiny seahorse. The basal ganglia are complex, made up of a number of tiny interconnected shapes. These three miniature, intricate structures enable us to feel and recognize emotions, store memories and control movement. If the amygdala isn't functioning, it's difficult to recognize emotion on people's faces. Without the hippocampus it's impossible to make and store new memories. For people with damage to this region, the world seems to be starting afresh every few minutes. If the basal ganglia aren't working, it's challenging to make coordinated actions, to control smooth movements and to learn new sequences of movement.

Below the subcortical structures, right at the base of the brain, is the cerebellum. It looks like a miniature brain, and this is what its name means (the word in Latin being the diminutive of *cerebrum*). It's intriguing because, although much smaller than the cortex, the cerebellum contains many more cells. All in all, the brain contains around 86 billion nerve cells (neurons), and 50 billion of these are found in the cerebellum. We don't know why this is the case. What we do know is that the cerebellum enables us to make smooth movements and controls balance and posture. Without it, it's difficult to move your hand without a tremor or touch your nose with your finger, for example.

As I held this beautifully intricate organ in my hands, I realized how much there is about ourselves that we don't know – how much of human thinking, feeling and doing we have to allocate among these tiny, complex forms and spaces.

The cortex and the subcortical structures make up what is known as the brain's grey matter, a mix of neuronal cell bodies, blood vessels, connections between neurons (synapses), unmyelinated axons and small cells called glia, which protect and support neurons. In that first dissection lesson, I learned – I *saw* – that these subcortical structures link up with one another and with the cortex. The links with the cortex are large tracts of what is called white matter: visible highways transporting vast quantities of information from the subcortical regions that process emotion, memory and movement, up to, for example, the prefrontal cortex, which incorporates this information into current decisions and future plans.

The prefrontal cortex is at the front of the brain, just behind the forehead. I knew, holding this large chunk of tissue in my hands, that this region, arguably more than any other, is what makes us human. The prefrontal cortex is bigger in humans than in any other species (relative to body size), and the cells it contains look different in humans from how they look in other animals. Decades of research have shown us that the prefrontal cortex is where decisions are made, where temptations are muted and where self-evaluation arises.

Together, these structures that make up the brain create each of us – the person, the self. The whole is infinitely more than the sum of the parts.

∞

These days, when we want to look at the brain, we're not restricted to dissecting post-mortem brains in the lab; we can also see inside the living, developing human brain, using the brain-scanning technology that has opened up unprecedented possibilities in neuroscience. I will come back to brain scanning in the next chapter. Suffice it to say here that things weren't always so easy. One early way of learning about human brain development involved slicing post-mortem brains into thousands of thin slivers and studying them under a microscope. I haven't worked with human brain tissue like this myself, but I work with people who do this kind of research and have observed them as they hunch over their microscopes. It is fiddly and arduous work, requiring meticulous dexterity and patience.

One of the first scientists to study the development of the human brain was Peter Huttenlocher (1931–2013). Huttenlocher was a paediatric neurologist working at Yale University and the University of Chicago from the 1960s onwards. He spent several decades exploring what the brain looks like in children of different ages, in both healthy children and those with intellectual disabilities.

Among much else, Huttenlocher was interested in understanding what brain tissue, as viewed under a microscope, looked like in children who suffered from seizures. (A condition that produces seizures was subsequently named after him and one of his colleagues as Alpers–Huttenlocher syndrome.) Another area he was interested in was the syndrome he called tuberous sclerosis, a rare condition in which mostly benign (non-cancerous) tumours grow throughout the body and the brain. Huttenlocher set up the first US paediatric clinic specifically for children with tuberous sclerosis. This condition sometimes causes developmental delay and learning difficulties, and Huttenlocher was interested in why this was the case. He was interested in how the brains of children with these and other developmental conditions differed from those of typically developing children.

To find out what was going on in these conditions, Huttenlocher had to look at slices of tissue from post-mortem brains under a microscope. This meant waiting for deliveries from local hospitals of brains from people who had died – infants, young children, and some adults for comparison – so that he could study slices of their brain tissue.

Each brain was delivered to Huttenlocher in a container of preserving fluid, from which he would lift it carefully on to his lab bench. Using a sharp knife, he would then cut the brain into large sections, separating the four lobes of the cortex and the subcortical regions

– those small structures deep inside the brain. His particular interest was in how the cells develop within the prefrontal cortex, the part of the brain involved in decision-making, planning and self-control, in comparison to the sensory regions such as the visual cortex at the back of the brain and the auditory cortex at the sides.

It was already known that damage to the frontal lobe has a more severe effect on personality and intelligence than damage to other parts of the brain. One famous case dates back to the mid-nineteenth century, around a hundred years before Huttenlocher started his work. This case, which featured in all the medical literature at the time and still appears in today's textbooks, was that of Phineas Gage. Gage was a railway worker in the American state of Vermont whose job was to prepare the ground for new railway tracks. On 13 September 1848, he was using a tamping iron to compact explosive powder into a deep hole in rock when the powder caught fire and exploded, shooting the entire 43-inch tamping iron through Gage's skull. It entered his head underneath his left eye socket, shooting right through the front of his brain and out through the top of his skull.

Phineas Gage with the tamping iron through his skull

Miraculously, Gage survived for twelve years after his injury, but his personality and his behaviour were, according to those who knew him, dramatically changed. While previously he had been apparently calm and well-liked, after the accident he was described by his doctor, John Harlow, as obstinate and capricious. This is what Harlow wrote about him while Gage was in his care (his work was published in 1868, after Gage had died):

> The equilibrium or balance, so to speak, between his intellectual faculties and animal propensities, seems to have been destroyed. He is fitful, irreverent, indulging at times in the grossest profanity (which was not previously his custom), manifesting but little deference for his fellows, impatient of restraint or advice when it conflicts with his desires, at times pertinaciously obstinate, yet capricious and vacillating, devising many plans of future operations, which are no sooner arranged than they are abandoned in turn for others appearing more feasible. A child in his intellectual capacity and manifestations, he has the animal passions of a strong man. Previous to his injury, although untrained in the schools, he possessed a well-balanced mind, and was looked upon by those who knew him as a shrewd, smart business man, very energetic and persistent in executing all his plans of operation. In this regard his mind was radically changed, so decidedly that his friends and acquaintances said he was 'no longer Gage'.

The case of Gage led many physicians, at the time and later, to conclude that the frontal lobe is the seat of personality, planning and self-control: Gage had lost his left frontal lobe and in consequence these aspects of his character had radically and irrevocably changed. More recently, this stark conclusion based on a single case has been questioned and disputed, particularly because there is no reliable description of Gage before his accident. Without this it's impossible to know how much he really had changed after the accident, and how much any change was due to the damage to this part of his brain.

Nevertheless, there were many subsequent cases of frontal lobe damage throughout the first half of the twentieth century, particularly during the two world wars. These cases all similarly pointed to damage to this brain region being detrimental for self-control, and leading to apparent dramatic changes in personality.

By the 1960s and 1970s, Huttenlocher and other scientists had inferred from Gage's and subsequent cases that the frontal lobes must be involved in the sense of self and high-level cognitive abilities such as intelligence, empathy, planning and self-control. This is what led Huttenlocher to concentrate on the prefrontal cortex in his work on child brains: he guessed this was where he would find differences in children with intellectual disabilities.

Once he had divided the brain into small sections, Huttenlocher sliced tiny slivers of tissue slowly and patiently from each brain region using a cryostat, an extremely sharp cutting instrument mounted inside a chamber kept at a very low temperature (about −20 degrees Celsius) in order to preserve the tissue being cut. Each sliver, no thicker than a sheet of cellophane, was then carefully fixed between two thin pieces of glass. Weeks of meticulous slicing and fixing would result in thousands of tissue samples from each brain region, which could then be studied under a microscope.

Huttenlocher was particularly interested in *synapses*, the tiny junctions between brain cells, or neurons, through which these cells communicate with one another (see diagram on p. 60). Through synapses, each neuron passes information to thousands of others. (I will talk about synapses in more detail in chapter 6.) Huttenlocher wanted to know about the numbers of synapses in different brain regions. To do this, he counted the number in each slice of brain tissue he was looking at under the microscope. In the 1960s and 1970s there was no automated way to do this – it had to be done by eye. This was difficult and tedious work that involved intense concentration for long periods of time. Huttenlocher spent many months counting millions of synapses, both at work and at home. Indeed, his daughter said after his death: 'He had all of these pictures of synapses in our house.'

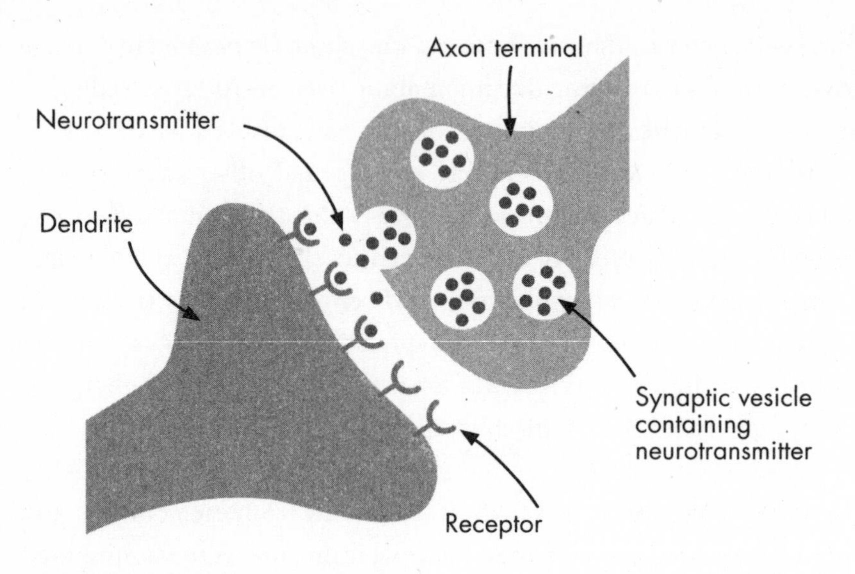

Detail of a synapse

Huttenlocher had set out to understand how the brains of children with intellectual disabilities differ from the brains of typically developing children; but, as he himself later remarked: 'Paradoxically, in our early studies, the findings in the normal population were more interesting than the abnormal population.' This was something of an understatement: Huttenlocher's findings astonished him and other scientists of the time.

Research on animal brains had already revealed how synapses develop. Early in brain development, even before birth, the branches that extend from each neuronal cell body (called *dendrites*), and the synapses attached to each neuron, start to multiply. This process, called *synaptogenesis*, continues for many months or even years – the exact amount of time depends on the region of the brain and the species of animal – producing an extremely large number of synapses, far in excess of what the brain will ever need. Indeed, the brains of young animals, or young children, contain vastly more synapses than the fully matured brains of adults.

This overproduction of connections is a universal developmental process – as far as we know, it happens in all brain areas and in all species of animal. So what happens to the excess synapses in the young brain, which have disappeared by adulthood? The reduction occurs via a process called *synaptic pruning*. Which synapses remain and which are eliminated depends at least in part on environmental experience. Synapses that are being used in a particular environment are retained and strengthened; synapses that are not being used are 'pruned' away.

The process of synaptic pruning is analogous to pruning a rose bush: weaker branches are pruned in order that the remaining branches can grow stronger. Another analogy is that of building a business. Most businesses start out by employing a limited number of people with the relevant skills. This makes economic sense, but there is an alternative option. You could, when you first set up your business, employ a very large number of people with many different skills, wait to see which skills turn out to be most useful in the prevailing business climate, and then retain the small fraction of people with those skills while laying off the majority of employees who don't have them. In the world of business, this would be a clearly inefficient and arguably unethical way to achieve optimal performance, and it doesn't often happen. But this is how the brain develops its connections. Early on, the brain generates many, many more synapses than it will ultimately need, and then gradually 'lays off' the synapses that are redundant. In this way, it ensures that all the necessary connections are made and retained.

An example of a process attributed to synaptic pruning is the loss of the ability to discriminate between certain sounds in language. The range of such sounds that an individual will be able to distinguish is determined by the sounds in that person's environment in their first twelve months of life. By the end of their first year, babies are no longer able to distinguish between sounds to which they are not exposed.

There are many sounds in languages that differ from our own, and to which we have not been exposed early on in life, that we cannot tell apart. Janet Werker, Renée Desjardins and their colleagues in Canada carried out a clever experiment in which they presented young babies, children and adults with two different sounds (such as 'da' and 'ba') and asked participants to indicate when they heard a difference between the sounds. The babies, who couldn't report the differences, were instead trained to turn their heads when they heard a different sound. Not surprisingly, participants from all age groups were able to hear the difference between 'da' and 'ba'. However, when the researchers presented the same participants with two very similar sounds from the Hindi language, the children and adults were unable to hear any difference between the two sounds – they sounded identical. Intriguingly, babies under one year *were* able to detect the difference. What this means is that it's usually difficult to develop a flawless accent in a second language that contains sounds that differ from your first language, unless you're exposed to it in the first year of life. The first year of life is an example of a *sensitive period* – in this case for sound categorization.

Can sounds to which we aren't exposed in the first year of life be relearned after the sensitive period is over? The answer seems to be yes – sometimes, and with effort, although this relearning is not equally successful in all people. Intriguingly, there are changes in brain activity that are associated with very subtle differences between foreign sounds that adults are not aware of. So the brain seems to detect differences between sounds that are impossible to hear. This result was discovered by Annette Karmiloff-Smith (1938–2016), a brilliant and pioneering developmental psychologist, and her colleagues at the University of London. It suggests that the brain retains the ability to detect subtle differences between sounds, and what is lost is the capacity to treat them as significant. This enables the most commonly encountered sounds to be given prominence, while other, less relevant sounds are filtered out.

For new sound learning to occur after the sensitive period, social interaction with other people seems to be essential. A study by Patricia Kuhl at the University of Washington in Seattle demonstrated that 9-month-old babies are able to learn new speech sounds to which they have not previously been exposed – as long as the new sounds come from a real person who interacts with the babies. Almost no learning occurred at this age from audio recordings or films of exactly the same sounds spoken by the same person for the same amount of time. Interaction with a real person was the key to learning.

Changes in the ability to perceive sounds over the first year of life are thought to be the result of synaptic pruning. The sound-processing synapses that are not being stimulated – because the language in the baby's natural environment does not contain certain sounds – are pruned away. Synaptic pruning thereby fine-tunes brain tissue so that it is most efficiently suited to the environment in which the child is growing up.

Until Huttenlocher's work, scientists had assumed that the human brain doesn't develop much after childhood. At birth, in all three regions of the human brain that Huttenlocher studied – the visual cortex, the auditory cortex and the prefrontal cortex – each neuron has more than 2,500 synapses. The number of these connections then increases rapidly in the first few months of life. In the visual cortex, the large region at the back of the brain that processes visual stimuli and enables us to see the world, the number of synapses peaks at around 8–12 months and then starts to decline again, levelling out at around 12 years. In the auditory cortex, the region involved in processing sounds and hearing spoken language, the peak occurs at around 3 months, and then declines until around age 12. This means that the numbers of synapses in these sensory regions don't change much after 12 years old, by which time they have already reached adult levels.

What Huttenlocher saw in the prefrontal cortex, however, told a different story. In this region, the number of synapses continues to increase throughout early childhood, and doesn't peak until age 3 years – later than in the auditory or visual cortex. After this, the number of synapses starts to decline, but gradually, continuing to decrease through the teenage years and not stabilizing until some point in the late teens.

Huttenlocher's findings indicated – for the first time – that certain parts of the brain don't stop developing in childhood, as was previously believed, but instead *continue to develop throughout childhood and adolescence.* Furthermore, his findings hinted at the possibility that environmental experience might play an important role in shaping the human prefrontal cortex in adolescence. He showed that synapses in this region of the brain are still being pruned away in adolescence, and we know that synaptic pruning depends partly on the individual's environment. Synapses that are required to process the particular stimuli in one's environment are maintained and strengthened, whereas those that are not needed are pruned away. The implication of this was that an adolescent's environment – culture, education, home and social life, hobbies, nutrition and exercise – might contribute to shaping their brain. It was a startlingly new notion.

Until Huttenlocher, all the previous data from neuroscience had pointed to the vast majority of brain development occurring early in life. This is because most previous work had focused on sensory regions (auditory and visual areas) of the brain in animals, which indeed do develop relatively early. It is difficult to overestimate the lasting impact of Huttenlocher's work, but initially scientists were sceptical about his findings. He was only able to study a total of twenty-one post-mortem brains from typically developing children and adults. Not many people donated their brains to science in his day, so it was not possible for him to obtain any more. What this means is that a lot of data points are missing in his work. Huttenlocher could see how many synapses there were in the prefrontal cortex of a brain at around age 6 and at 16 years, but he had no brains to study from between those two ages. Those data were simply missing, and he could only guess at what might happen between these ages by drawing an imaginary line between the data points.

∞

When I read Huttenlocher's papers describing the development of the human brain, I am often struck by a sense of sadness and poignancy. His work has taught us so much about the brain, suggesting for the first time that some regions go on developing beyond childhood. And yet it relied entirely on the deaths of individuals, old and young. I imagine there must have been a bittersweet feeling when a young brain was delivered to his laboratory. He was desperately in need of post-mortem brains from children and young people to fill out his painstakingly compiled picture of brain development; but, at the same time, knowing he was holding the brain of a child must surely have been distressing. Scientific knowledge about brain development owes a great deal to these deceased people and the generosity with which they, or their families, donated their brains to research.

Huttenlocher's discoveries paved the way for a whole field of research on the developing human brain – but, until relatively recently, there were limited ways of pursuing it. Opportunities to study slices of post-mortem human brain tissue, as we have just seen, were rare. So if scientists wanted to understand which parts of the brain were involved in which cognitive processes – memory, attention, language or emotion, for example – they often studied living patients who had sustained damage to part of the brain following a head injury or stroke. The scientist would be given information about which part of the brain was damaged, and they would work to identify what cognitive or sensory process was impaired as a result. They would then be able to conclude that a particular brain region was responsible for that cognitive process.

There are many well-known examples of brain-damaged patients and what they have taught scientists about how the brain works. Such cases have been brilliantly described in books by Oliver Sacks and, more recently, by Paul Broks.

A famous early case from the United States was that of Henry Gustav Molaison (1926–2008). In the medical and psychological literature he was – and still is – known by his initials, HM, and many hundreds of research papers have been written about him. HM suffered from severe epilepsy, which was resistant to treatment. His doctor suspected that his epileptic seizures originated in his medial temporal lobes and surrounding areas, including the hippo-campus, that tiny region deep in the centre of the brain. As a result, his doctor – following common practice in the mid-twentieth century – decided to remove a large portion of his medial temporal lobes in order to cure the epilepsy. At the age of 27, in 1953, HM underwent neurosurgery that removed most of his hippocampus, as well as removing or damaging other nearby regions of his brain.

The surgery did indeed seem to improve the epilepsy: HM's seizures were much reduced. However, his doctors soon noticed that it had left HM with severe memory problems. He could not form new memories – a condition called *anterograde* amnesia. In addition, he had moderate *retrograde* amnesia: no memory of most events in the two-year period before the surgery, and no memory of some events up to eleven years before. What particularly interested psychologists at the time was that, despite not being able to lay down new memories of people, facts or events, HM could still learn new motor skills – that is, he still had what is called *procedural* memory. Procedural memory, which is used in learning and retaining motor skills such as riding a bike, kicking a football or playing the piano, involves different regions of the brain, the basal ganglia and the cerebellum, which were undamaged by the surgery HM had undergone.

HM's case was studied by hundreds of neurologists and psychologists from 1957, when the first paper about his poor memory was published by Brenda Milner, the psychologist who studied HM for decades, and William Beecher Scoville, the neurosurgeon who had operated on HM. The case had a substantial impact on our understanding of memory, for it was the first detailed case study of amnesia. In addition, the doctors knew which parts of the brain had been damaged or removed by the surgery, and this supported the idea that particular parts of the brain were responsible for particular psychological functions. In this case, it was concluded that an intact hippocampus is crucial for certain forms of memory. In addition, the fact that HM still had some older memories, had retained words and was able to learn new motor skills suggested that these different forms of memory reside in different brain regions.

Much of the scientific debate that has surrounded the case of HM has focused on the precision of the surgery: brain scans in the 1990s revealed that a larger portion of HM's brain had been damaged than was previously assumed. Nevertheless, HM's contribution to the branch of science called neuropsychology should not be underestimated. After his death in 2008, his brain was donated to science; it has since been cut into thousands of thin slices and studied under the microscope by a group of scientists at the University of California in San Diego, who have also created a digital three-dimensional reconstruction of HM's brain.

Studying patients with damage to the hippocampus has taught neuropsychologists that this region of the brain is critically involved in transferring experiences from short-term to long-term memory. Imagine if you were unable to form new memories: every day you would have new experiences, but none of them would be retained in memory or become part of who you are; all new experiences, encounters with new people, new facts and personal events would vanish. You'd build up no autobiographical memories and you'd have no sense of the places you'd been, the people you'd encountered or the day-to-day experiences that had happened to you. And, as we've already seen in chapter 2, memories are important in forming a sense of who you are, a sense of self.

From Phineas Gage onwards, we have learned a great deal about the human brain by studying the effects of brain damage on cognitive function. We have also learned that the picture is complicated. Damage often affects multiple brain regions, so we can't be sure which region is responsible for which function. Conversely, damage to one region often affects several different cognitive functions; we might only test one, but we can't rule out the possibility that others are also affected. Other characteristics of this kind of research make it tricky, too. The patient might be in pain, tired or non-cooperative, so that it's often hard to tell whether it's the brain damage that's causing the poor cognitive performance. There's also a degree of luck involved – neuropsychologists have to wait for patients to show up at hospital and study whomever happens to come in with a neat enough lesion to investigate.

What was needed was a more straightforward way to find out which parts of the brain are involved in which cognitive processes, and in healthy brains that had not been damaged. That's where magnetic resonance imaging (MRI) – a type of brain-scanning technique – comes in. MRI, which was first used to study the human brain in the late 1980s, has given us unprecedented access to information about this most complex of organs. Because it allows us to study the living, working brain in healthy individuals of all ages, it opens up huge vistas of research that were firmly closed off to earlier scientists, however dedicated. MRI scanning has transformed modern neuroscience. And it has enabled us to home in on exactly what is going on between Huttenlocher's 6-year-old and 16-year-old – to track the development of the living adolescent brain.

5

Inside the living brain

THE FIRST TIME I had a brain scan was in 1997. I was in London, doing my PhD at UCL, and I'd volunteered for a colleague's study. When we entered the scanner room, the first thing I thought was: 'What? I'm supposed to get *into* that machine?'

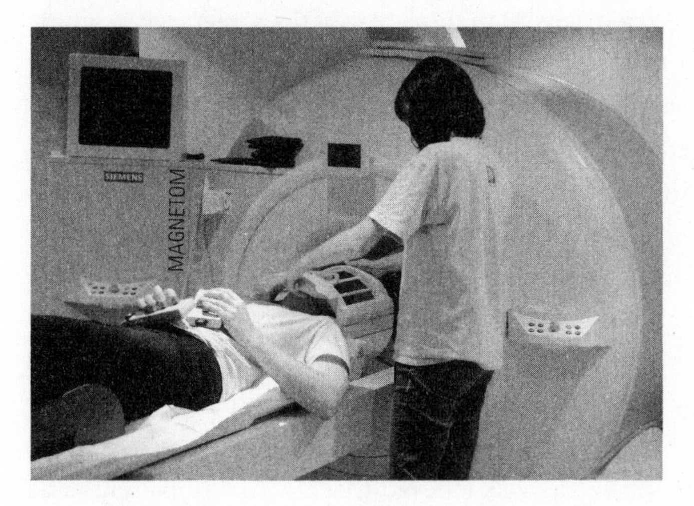

An MRI scanner

An MRI scanner is essentially a long, narrow tube. It looked worryingly narrow to me. I had to lie on a table that would be moved backwards into the scanner with me on it. My colleague, Geraint Rees, and the radiographer secured my head with foam cushions to keep it still while I was in the scanner, and fitted me with earplugs to protect my ears from the noise, which I was told would sound like techno music when I was inside.

Once the table on which I was lying had moved slowly into the scanner, the researchers turned out the main lights and left the room, shutting the heavy door behind them. I was beginning to feel distinctly uneasy. What if something went wrong? Could I escape from this narrow tube if I had to? Would I be able to contact the researchers in the room next door? What if the communication buzzer stopped working?

Before I had much time to ponder these questions, the experiment began. The scanner vibrated, the techno sounds started, and words and pictures were flashed on a computer screen in front of my eyes. The experiment was looking at how the brain enables us to pay attention to some things going on around us and not to others, and I had to answer questions about a word in the centre of the screen, while ignoring distracting pictures, which popped up in the periphery. The scanner, meanwhile, was taking a three-dimensional picture of my whole brain every three seconds, and recording my brain activity as I tried to follow the instructions. The task was quite a challenge, and as I concentrated on answering the questions, one after the other, I forgot about my cramped surroundings for forty-five minutes, by which time the functional scan (fMRI for short) was over.

I then had a structural scan (very simply, a still snapshot of my brain), during which I lay motionless with my eyes shut while the MRI scanner vibrated and made a noise that sounded like a different kind of techno music. By then I was relieved that the functional part was over and I was able to lie there doing nothing for ten minutes. Instead of recording my brain in action, now the scanner was taking a detailed photograph of the structure of my brain, in three dimensions.

After the scanner had completed its work and I had emerged, Geraint showed me the structural MRI scan of my brain on a large computer screen, using the computer mouse to zoom around the inside of my brain. I was amazed by the detail and clarity of the images. You could see the folded layers of the surface of the cortex, the cerebellum, the subcortical regions such as the amygdala – and, more surprisingly, my tongue, my nose and my eyeballs. It was even possible to recognize a vertical slice of the scan as me because it had captured my profile.

Since then, I've volunteered for dozens of research MRI scans, and also had clinical MRI scans for headaches, so lying in a scanner for an hour or so is now second nature. I feel none of the anxiety I felt that first time, and working in the field for over twenty years has given me confidence about the safety of the process. The only time it's not safe is if you have any metal object in your body or your clothes when you enter the scanner. A coin or a hair clip is magnetic, and the scanner is, essentially, a huge magnet. Anyone with metal inside their body – such as a surgical implant or a pacemaker – is not allowed inside the scanner. Even dental braces and fillings can cause some distortion of the images of the brain. As well as messing up the quality of the brain images, a loose metal object might be attracted towards the magnet and cause damage to the scanner – or, worse, to the person inside it.

MRI uses a magnetic field to produce high-quality three-dimensional images of the brain and the body. A large cylindrical magnet creates a magnetic field around the person's head, and a magnetic pulse is sent through this field. Different structures in the brain, for example white matter and grey matter, blood vessels, fluid and bone, have different magnetic properties and therefore appear distinctly different in the MRI image. Sensors inside the scanner record signals from the different brain structures and a computer uses the information to construct a three-dimensional image by placing many two-dimensional images on top of one another. Using MRI, scientists can construct a picture of both surface and deep brain structures in great anatomical detail. In fact, MRI scans look a bit like a three-dimensional X-ray photograph of the brain. The image of the brain created by an MRI scan contains hundreds of thousands of small (typically 1mm³) cubes called *voxels* (you can think of these as a 3D version of the pixels used to measure the definition of a digital photo or computer screen), each of which is classified by a computer as either white or grey matter.

Techniques have been developed that enable scientists to use MRI not only to produce a reconstruction of the brain, but also to observe the brain at work. This is functional MRI or fMRI – as in the experiment I described at the beginning of this chapter. Imagine you're talking to a friend. When you do this, many billions of the neurons in your brain are activated – to enable you to recognize your friend's identity and her facial expressions, listen to what she is saying, think about your response, produce speech and laughter, and so on. When your neurons become active they require energy, and this is provided in the form of a constant supply of glucose and oxygen carried in your blood. In fact, your brain uses up a fifth of your body's energy – 20 per cent of the calories you consume are used to fuel the brain. The fMRI scanner detects the levels of oxygen in the blood, because oxygen has magnetic properties. By measuring the amount of oxygenated blood that is sent to particular regions in the brain, fMRI can make 'films' of changes in brain activity as volunteers carry out a task such as reading words, looking at pictures or hearing sounds.

Over the past two decades or so, we have learned which parts of the brain are involved in pretty much every conceivable cognitive and emotional process – or at least, everything it's possible to do inside the confines of an MRI scanner. We know which brain regions are involved in basic sensory processes such as seeing objects, hearing sounds, feeling touch and moving. We also know which regions are involved in more complex sensory processes such as recognizing an actor's face or your partner's voice, perceiving emotions such as fear and anger, writing your signature and imagining playing the piano. We even know which regions are involved in highly complex cognitive processes, such as empathy, thinking about other people's minds, feeling complicated emotions such as guilt and embarrassment, and falling in love.

As well as all this, MRI scans have revealed a great deal about the developing brain. Just as Huttenlocher discovered through his diligent examinations of post-mortem brain slices, structural MRI scans have shown that one of the brain regions that undergoes the most striking and protracted change during adolescence and early adulthood is the prefrontal cortex. Remember that this is the part of the brain at the very front, which is relatively bigger in humans than in other animals, and is involved in a wide variety of high-level cognitive and executive functions, including decision-making, planning, inhibition of inappropriate or risk-taking behaviour, social interaction – understanding other people – and self-awareness.

Other regions in the cortex, such as the temporal cortex, which is involved in language, memory and social understanding, and the parietal cortex, which is involved in planning movements, spatial navigation and multi-sensory processing, also undergo protracted development in adolescence. The important overall finding is that adolescents' brains are physically different from younger children's brains and from adults' brains. In terms of brain development, adolescence is a distinct stage.

∞

In chapter 2, I described a study we did in my lab on introspection – our ability to judge our own performance on a task. One of our collaborators on that project, Steve Fleming, had already carried out a study to identify the part of the brain that is related to introspective ability in adults. In this study, a group of adults carried out the same introspection task that we used, and also had a structural MRI scan. Steve and his colleagues then looked at the relationship between introspection ability and grey matter volume across the brain. Their results revealed that introspection ability is associated with the size of the right rostrolateral prefrontal cortex: the better you are at introspecting, the larger this part of the brain is. This is a region at the very front of the brain, on the right side. The size of this region – how much grey matter it contains – wasn't associated with how accurately people distinguished between the two pictures, but with the participants' ability to judge the accuracy of their performance. People with more grey matter in this area tended to be better at judging their own performance.

We don't know why this is. It could be that grey matter volume in this area determines someone's introspective ability. On the other hand, an equally plausible explanation is that practice at introspection increases the size of the right rostrolateral prefrontal cortex. The results of most neuroimaging studies are simply correlations, showing an association between the size of, or activity in, a certain brain region and something else: a state of mind, a characteristic of the person, or such-like. It's not possible to know what causes what. The size of a brain region might cause you to be better at something – introspection in this example – or practising that something might increase the size of the particular brain region. Or both could be true.

We *do* know that practising a task leads to changes in brain structure. A famous example concerns London's black-cab drivers. In order to become a black-cab driver, you need to pass a test called The Knowledge, which involves memorizing the roadmap of London, including around 25,000 streets and thousands of landmarks – a task that usually takes at least two years to complete. Eleanor Maguire and her colleagues at the Wellcome Trust Centre for Neuroimaging in London scanned the brains of experienced cab drivers and compared the structure of the (male) cab drivers' brains with the brains of men around the same age who were not cab drivers. One region of the brain was of particular interest to them: the hippocampus. This is the small, seahorse-shaped structure in the middle of the brain that I discussed in relation to memory and the case of HM in chapter 4. As well as storing 'episodic' memories of past events, and laying down new episodic memories, the hippocampus is also involved in spatial navigation – remembering how to get from A to B.

Maguire's studies showed that there were intriguing differences in hippocampus size between the London cab drivers and the other men: the posterior (back part of the) hippocampus was significantly larger in the cab drivers. And, importantly, the longer the person had been driving taxis, the larger this part of their hippocampus was. This finding – that hippocampus volume was related to the time the person had been driving cabs – indicates that hippocampal size depends on how much someone has used their spatial memory, rather than the other way round. That is, it suggests that it wasn't the case that men who have a large posterior hippocampus have better spatial navigation skills and are therefore more likely to become cab drivers. The findings from the cab-driver studies suggest that the more you use your spatial navigation, the bigger your posterior hippocampus becomes.

So it's possible that the rostrolateral prefrontal cortex may increase in size as we practise introspection – a skill that, as we saw in chapter 2, improves in adolescence.

Another part of the prefrontal cortex that plays a key role in the sense of self is the lower portion of the dorsomedial prefrontal cortex (dmPFC). In brain-imaging experiments, this region becomes active when volunteers think about character traits and attributes that describe themselves. One of the first brain-imaging experiments on the self was led by Kevin Ochsner at Columbia University in New York City. Ochsner and his colleagues found that the dmPFC was activated when volunteers were asked to think about how much personality-trait adjectives, such as 'trustworthy' or 'shy', described themselves compared with how much these words described someone else close to them (their mother or their partner, for example) or the average person. You can imagine doing this task by thinking about how different adjectives describe you compared with your mother, say. How *reliable* are you compared with your mother? How *assertive*? How *anxious*?

As with all brain regions, the dmPFC does not have only one function; it's involved in many different aspects of cognitive processing.* However, its role in thinking about ourselves is particularly interesting to us here because, as noted above, the prefrontal cortex is a brain region that undergoes substantial development during adolescence.

Developmental neuroimaging studies carried out over the past decade or so have compared self-reflection in different age groups. One of the first developmental neuroimaging studies on the self was published in 2007 by Jennifer Pfeifer and her colleagues at the University of California in Los Angeles. Children aged between 9 and 11 years, and adults aged between 23 and 32 years, were scanned while they judged whether phrases such as 'I like to read just for fun' described either themselves or Harry Potter (Harry was chosen because he was a universally familiar but fictional other).

Pfeifer and her colleagues found that in the children, the dmPFC was activated to a greater extent when thinking about the self than when thinking about Harry. This wasn't true for adults. Interestingly, in the adults, a different region – the lateral temporal cortex – was activated more than it was in the children when thinking about the self. The lateral temporal cortex plays a role in memory retrieval, and so the researchers suggested that adults might use stored self-knowledge when performing the task more than children do, while children might rely less on stored knowledge about the self, and more on self-reflection using the dmPFC.

* Sensory regions of the brain have more than one function. The visual cortex, which makes sense of visual information and allows us to see the world around us, starts responding to touch in people who are blind, for example. Studies have shown that even in sighted people the visual cortex starts to respond to sounds if a particular sound is heard many times in conjunction with a particular visual stimulus. The visual cortex is also activated when we simply think about visual objects – if you imagine a pineapple, for example.

These suggestions recall the speculation about developmental change in the pattern of activity during thinking about intentions, which I discussed in chapter 2. It's important to reiterate that we can't be sure of the cause – the relevant brain regions are also involved in other processes too, and it's impossible from neuroimaging experiments to be sure what psychological process is causing a brain region to become activated. Also, we cannot make inferences about what goes on in the brain during this task in adolescence, as adolescents weren't included in this study. Nevertheless, the results do suggest that children and adults might use different mental approaches when making judgements about their own character traits.

When MRI was first used to scan people's brains, all studies involved scanning adult volunteers. In the past two decades, neuroscientists have started to examine the living human brain at all ages in order to track developmental changes in the brain's structure and also how it functions across the lifespan. Many laboratories around the world do research in this area, and we now have a rich and detailed picture of how the living human brain develops. This picture has changed the way we think about human brain development, by revealing that development does not stop in childhood but continues throughout adolescence and even into adulthood. Huttenlocher was right – and not just about the prefrontal cortex; it turns out that many regions of the human brain continue to develop well after childhood.

MRI scans do much more than confirm some of Huttenlocher's key findings. In fact, they offer us a completely different way of looking at the brain from the post-mortem brain slices that Huttenlocher studied. If you look at slices of brain tissue under a microscope you can see cells and synapses. This isn't possible with MRI – although MRI scans are beautifully detailed, their resolution is much less precise than the image of a brain slice under a microscope. As I have mentioned, what we see in MRI scans are computerized pictures of the brain made up of thousands of those little cubes called *voxels*, each one of which is classified as either white matter or grey matter. Each MRI voxel of white matter contains thousands of *axons*, the long fibres along which electrical signals pass as brain cells – neurons – communicate with one another (see the illustration in chapter 6). We can't see this detail in an MRI scan, but we know that each voxel of grey matter contains tens of thousands of neurons and millions of connections (synapses). And in the last twenty years we've learned much more about what happens to grey matter and white matter in the human brain as it develops. It's this research, and the exciting discoveries it has produced about exactly *how* and *when* the brain changes, that will be the focus of the next chapter.

6

The ever-plastic brain

ALTHOUGH HUTTENLOCHER'S WORK IN the 1970s had provided some evidence about the development of the human brain, and MRI technology emerged in the late 1980s, when I was an undergraduate in the mid-1990s most of what was known about brain development still came from studies on animal brains. The textbooks I used then contained very little information about *human* brain development and, where they did, they implied that most of the changes occur very early in life with little happening after childhood. Here's an example, from a passage discussing myelination, the process by which axons are coated with myelin throughout development:

> It is one of the later stages of maturation, beginning usually late in embryonic life or in early postnatal life after the projection neurons are well in place, and continuing for considerable periods of time (into childhood, in the case of humans).

What have we learned since then? The answer is: an enormous amount, thanks largely to the advances in brain-imaging technologies outlined in the previous chapter. The most reliable and informative studies of human brain development are those that involve scanning hundreds or even thousands of the same individuals at multiple points throughout their lives – these are called *longitudinal* studies.

The first study of this kind was led by Jay Giedd at the National Institute of Mental Health (NIMH) in Bethesda in the United States.

Giedd was a true pioneer, one of the first scientists to study the developing human brain using MRI. In the early 1990s, with colleagues including Judith Rapoport, Giedd began a very large longitudinal study in which he scanned the brains of children, adolescents and adults as they got older. The study continued for over twenty years, during which hundreds of participants between the ages of 4 and 88 were scanned (it's very difficult to scan children younger than 4 years because it is almost impossible to get them to keep completely still inside the scanner, and any movement blurs the images). Each participant came in for scans every two years and had between two and seven scans, so the researchers were able to collect many thousands of brain scans overall, and multiple scans from each person.

This study yielded the first (and still the largest) dataset of the developing human brain. It has revealed a great deal about how our brains change throughout life as we grow and develop. Since Giedd began his work, multiple developmental MRI studies have shown that cortical grey matter volume decreases and white matter increases during adolescence. A 2016 study by Christian Tamnes and Kate Mills, along with many colleagues from different universities, analysed longitudinal data from 391 individuals aged between 8 and 30 years from four large cohorts in three countries: the United States, the Netherlands and Norway. One of the American cohorts was the NIMH sample described above. In all four groups of participants, cortical grey matter volume increased during childhood, reaching its peak in late childhood, and then declined across adolescence (see the graph opposite). There were differences between different parts of the brain, and again these were remarkably consistent across the four samples. The analysis revealed decreases in grey matter volume across the cortex throughout adolescence, with the largest decreases occurring in the prefrontal, parietal and temporal cortices. Why does this happen? I will come back to this in a moment.

∞

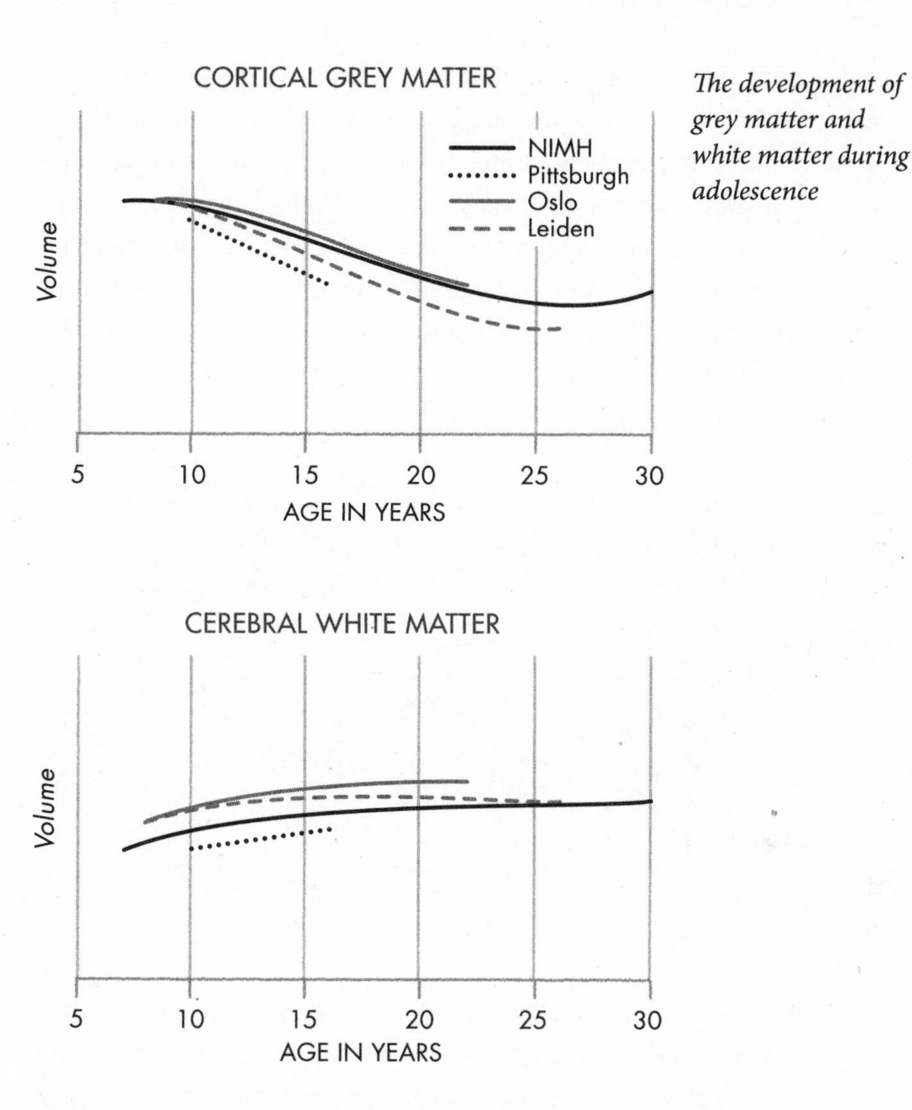

The development of grey matter and white matter during adolescence

In order to understand what MRI scans are able to show us about the brain, we need to look in a little more detail at the anatomy of the brain. The adult human brain weighs about 1.4kg (about 3lb) and, as I explained earlier, contains about 86 billion brain cells, called *neurons* (see diagram overleaf).

That's a big number – 86 billion – almost as big as the number of stars in our galaxy. But what really matters, in terms of the function of the brain, is the complexity of the 'wiring' between the neurons. Each neuron has a cell body with many short branches, called *dendrites*, around it, and one long, thin fibre, called an *axon*, which connects and communicates with the dendrites and cell bodies of other neurons across junctions called *synapses*. Some estimates suggest that there are as many as one million billion synaptic connections between neurons in the adult brain – that's a quadrillion synapses. A staggering number.

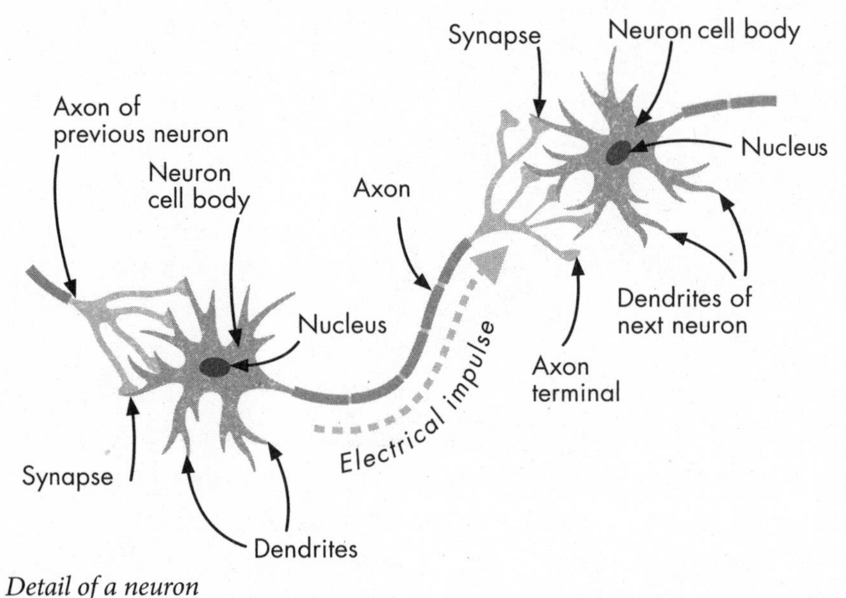

Detail of a neuron

Whenever you perceive, or experience, or do anything – when you recognize a friend, feel angry, fall in love, have a conversation, do sums, learn a new word, go for a walk, switch on your phone – millions of neurons in your brain are communicating with each other through hundreds of millions of synapses. How do they do this?

When a neuron is activated it fires an electrical impulse, called an *action potential*. Each impulse is a tiny pulse of electricity (about one-tenth of a volt), lasting about one-thousandth of a second, which is generated by the cell body and propelled down to the tip or *terminal* of the axon, where it causes chemicals called *neurotransmitters*, such as glutamate, serotonin and dopamine, to be released (see diagram on p. 60). The neurotransmitters move across the tiny gap of the synapse and are taken up by specialized protein molecules called *receptors* in the membrane covering the dendrites and cell body of the next neuron. Some of these receptors trigger electrical reactions in the next neuron and if those reactions are big enough, they produce a new impulse, which shoots down the axon of that neuron. And so on.

Impulses rushing along little fibres, from neuron to neuron: this is the 'language' of the brain. It is how neurons communicate with each other, and it is how the brain does its work. Seeing, thinking, speaking, acting – these are all the results of patterns of impulses transmitted through circuits of millions of nerve cells.

In order to make decisions and to act in real time, in the real world, neurons need to be able to communicate with each other quickly. A thin axon with no extra insulation around it conducts impulses as slowly as 1 metre per second. That might be good enough if you are small and slow, like a snail, but it's hopeless for big, rapidly moving animals – such as human beings. There are axons that run all the way down from the cerebral cortex of your brain to the base of your spinal cord – a metre or so in an adult human. Imagine how difficult it would be to control your legs if it took more than a second for every impulse to travel from brain to muscle.

Evolution has done its best to improve the inferior hardware. First, thick axons are faster than thin ones. However, the problem with very large axons is that they take up more space. A cleverer trick is to improve the insulation. Vertebrate animals have evolved mechanisms for wrapping a sheet of fatty material called *myelin* round and round the axon. This process, called *myelination*, speeds up the spread of voltage along the interior of the axon. The impulse still needs to be boosted, but in a myelinated axon this happens only at little gaps in the myelin sheath, called *nodes of Ranvier*, every millimetre or so along the axon. Relatively large myelinated fibres in human beings, such as those from the cortex to the spinal cord, can conduct impulses faster than 100 metres per second.

When axons first grow out from neuron cell bodies, very early in development, they have no myelin. Many of them, such as certain sensory axons that transmit information about pain and temperature from the skin, remain without myelin throughout life. But most long axons become myelinated at some point during development. Myelination is a very gradual process, beginning in the second trimester of foetal life and continuing during the rest of pregnancy and infancy, and throughout childhood, adolescence and early adulthood. The addition of myelin speeds up the conduction of impulses and this presumably facilitates the rapid processing of information in the brain, as well as the speed and accuracy of decisions and movements. Thus, the thickening of axons and the addition of myelin both improve the brain's capacity to transmit information. Reactions to events in the outside world can be faster.

The neurons of the cortex are packed together at the surface of the cerebral hemispheres, with a total thickness of 3mm or so. This layer looks dark in a slice of brain (taken during a post-mortem examination, for example). So it is called 'grey matter'. As noted in chapter 4, the grey matter contains not just the cell bodies of neurons but their dendrites, their own axons and other axons running into the grey matter from other regions, all the synapses, and also blood vessels and small protective and supportive cells called glia.

The huge numbers of axons that run into the grey matter, sending information to the neurons, and that run out of the cortex, connecting to other areas of the brain and spinal cord, lie underneath the grey matter, making up much of the volume of the cerebral hemisphere. Myelin reflects light, so it appears white in fresh sections of the brain. That's why the mass of axons within the brain is called 'white matter'.

An MRI machine, which detects water and fat in tissues of the body, can pick up the difference between grey matter and white matter.

The computer wizardry that turns the signals into a beautiful brain scan can make them any colour that the neurologist or the research scientist wants. But it's quite common for brain scan pictures to show the grey matter as grey and the white matter as white, to make the scans look more like real slices of brain.

The whole brain gets bigger for a long period after birth, although no (or very few) new neurons are made in the cerebral cortex after birth. The brain increases in volume fourfold between birth and age 6, and is then 90 per cent of adult size. MRI scans can be used to measure the growth and development of the brain (without having to take it out and chop it up).

What's most dramatic is the enlargement of the white matter. There is a steady increase in white matter volume in several brain regions during childhood and adolescence. This could be due to an increase in the actual number of axons, thickening of individual axons or the addition of myelin. All three probably contribute. And they all increase the speed or richness of connections within the brain. White matter volume stops increasing and finally levels off at some point in the thirties or early forties in humans.

MRI studies have shown that the volume of grey matter in various cortical regions increases during childhood and peaks in late childhood; it then decreases substantially during adolescence, with some regions of the cortex losing about 17 per cent of grey matter volume between late childhood and early adulthood. This decline levels off at some point in early adulthood, between the mid-twenties and thirties.

A loss of grey matter might sound like a bad thing. But it is very unlikely that the brain is degenerating and on the decline during adolescence. The gradual decrease in thickness of grey matter after childhood doesn't mean that neurons are dying. The decrease in grey matter during adolescence probably reflects a number of important neurodevelopmental processes, which help to refine and mould the maturing brain. To understand what these processes are, we need to know what's happening to the cells and the synapses during development, and this is something MRI can't tell us, because the resolution of the MRI scan isn't high enough. The precision of MRI is greater than other types of brain scanning, but still low compared with studying brain cells under a microscope. So, to understand what's happening with the cells and synapses in grey matter during development, we need to turn back to Huttenlocher's methodology and the study of post-mortem brain tissue observed under the microscope.

One possible neurodevelopmental process that may underlie the changes in grey matter volume seen in MRI scans is the change in the number of synapses and dendrites. In chapter 4, I described how studies of post-mortem brain slices have shown that synapses start to multiply from early on in a foetus's life, in a process called *synaptogenesis.* This continues throughout early development and results in a vastly excessive number of synapses, so that by the time a child is 1 year old, his or her brain is estimated to contain about twice as many synapses as an adult brain. Which synapses remain and which are eliminated depends at least in part on environmental experience; synapses that are not used are eliminated in the process called synaptic pruning. So the increase in grey matter during childhood might correspond (in part) to an increase in the number of synapses; conversely, the decrease in grey matter during adolescence and early adulthood might be due to synaptic pruning.

We know from Peter Huttenlocher's work in the 1960s and 1970s that synapses in the prefrontal cortex increase in number during childhood and decrease during adolescence; and a 2011 study showed the same pattern of results. Zdravko Petanjek and his collaborators at the University of Zagreb in Croatia studied thirty-two post-mortem brains from newborn to age 91. They found that the number of synapses in the prefrontal cortex is highest during childhood, starts to decline around puberty, continues to decrease throughout adolescence and the twenties, and eventually stabilizes in the early to mid-thirties. This pattern of synaptic development is similar to the pattern of grey matter development in the prefrontal cortex seen in MRI studies.

However, synapses are tiny, and synaptic pruning alone cannot account for the 17 per cent decline in cortical grey matter volume across adolescence. The second element in the explanation for the decline in grey matter during adolescence is the increase in white matter. As axons become myelinated, they change in appearance from grey to white. So, as white matter increases, there is a concomitant loss of grey matter, as brain tissue becomes more white and less grey.

People sometimes ask me when the brain *stops* developing – when does it 'become adult'? This question might seem highly relevant to issues such as the age of consent, voting threshold, criminal responsibility, alcohol consumption, joining the army and so on. It might be quite useful if brain research could help us to inform these difficult and complex issues, and to pinpoint a precise age at which the brain suddenly becomes 'adult'. But it's not that simple. Different brain regions develop at different rates and cease developing – the respective amounts of grey matter and white matter levelling off and then staying more or less the same for several decades – at different ages. The precise age at which an individual's brain stops developing probably depends on multiple factors, including genetic and environmental influences. Thus it varies from person to person. It might be more useful to consider the age at which brain development ceases as a broad *age range* rather than a specific number of years. The scientific studies suggest that in most people the brain will have stopped developing by the forties.

∞

Many studies have confirmed that one of the brain regions that shows the most striking and prolonged changes is the prefrontal cortex, which, as we have already seen, is involved in a variety of cognitive functions, including decision-making, planning, self-control, social interaction and self-awareness.

Studies have shown that, when the prefrontal cortex is damaged, a variety of cognitive functions are seriously affected. In particular, so-called executive functions appear to rely on the functioning of the prefrontal cortex. Executive functions enable us to plan and coordinate our decisions and actions, and to exert mental flexibility and self-control. As shown in the case of Phineas Gage, patients with prefrontal cortex damage find it difficult to plan – whether it's planning what they're going to do today, next week or next year.

In the laboratory, this has been tested in the 'Shopping Task', devised by Paul Burgess and Tim Shallice at UCL. In this task, patients with damage to the prefrontal cortex were given a shopping list of items that had to be bought from different shops in Lamb's Conduit Street, a small and rather beautiful pedestrian street in London, just around the corner from the National Hospital for Neurology and Neurosurgery, where the patients were being treated.* They were instructed to purchase all the items on the list within a set amount of time and by going into as few shops as possible. The most efficient way to do this would be to group the items on the shopping list according to which shop they could be found in – all toiletry items from the chemist, groceries from the supermarket, pens from the stationers and so on – and enter each shop only once.

Healthy participants are able to complete this task fairly easily. However, patients with prefrontal damage who took part in this study performed pretty badly. Instead of planning which shops to visit and in which order, they shopped in a haphazard way, entering the same shop several times, missing items on the list, and going back for things they had already bought (see the diagram opposite).

Patients with prefrontal cortex damage also find it difficult to inhibit behaviour that's rude or inappropriate. In everyday life, this can come across as seeming insensitive, saying or doing things that are

* It's also around the corner from my office, and my colleagues and I often wander there for lunch.

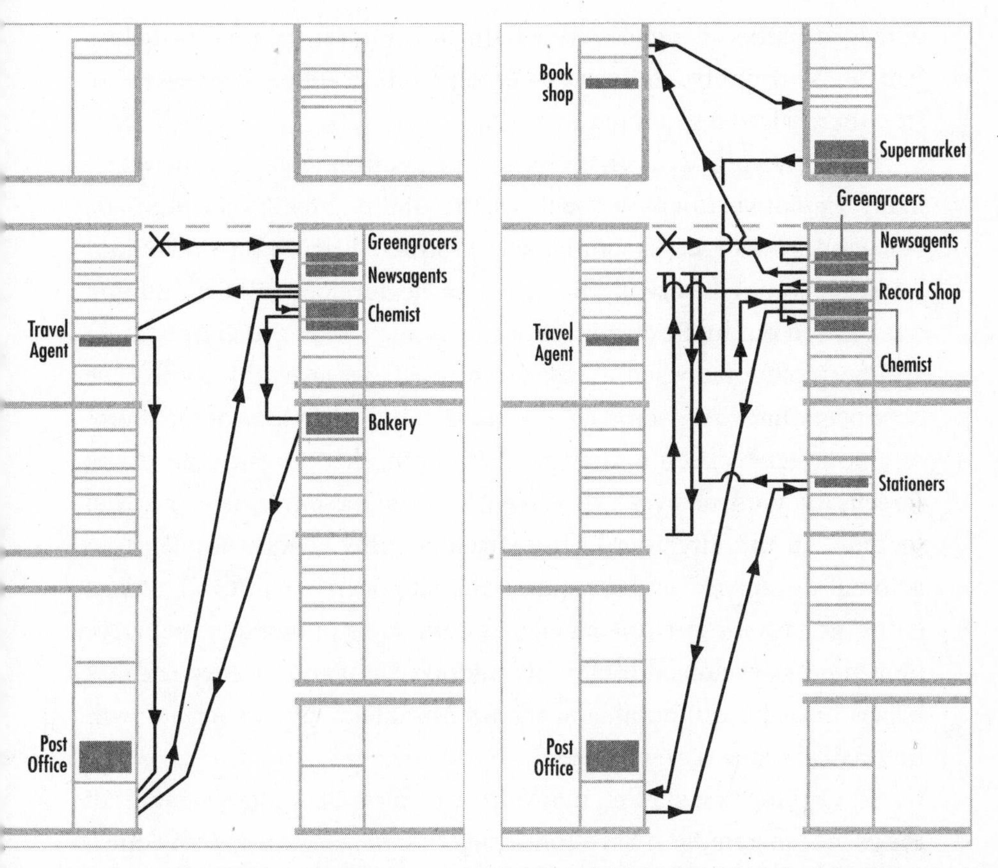

The Shopping Task: routes taken by a healthy person and by a patient with damage to the prefrontal cortex

socially inappropriate or impulsive. Patients with injury to this region might lose their temper quickly, show emotional outbursts at inappropriate moments or spend more money than they have, for example. In the lab, this impulsivity can be assessed in experiments in which participants are asked to press a button whenever a letter appears on a computer screen, *except* when the letter is an X, in which case they have to suppress their impulse to press the button. This is called a *go/no-go* task and is surprisingly difficult even for those with undamaged brains – we all find it hard to inhibit a habitual response. Patients with prefrontal cortex damage find it almost impossible to stop themselves responding to the X.

Cognitive processes that rely on the prefrontal cortex, including many executive functions such as the ability to inhibit automatic behaviour, undergo substantial and protracted development in adolescence. One of the first studies to assess the development of inhibition was carried out by Beatriz Luna at the University of Pittsburgh. Luna and her colleagues used a clever task that measures the control we have over our eye movements. If you ask someone to look at the centre of a computer screen and then flash a stimulus on one side of the screen, the person's eyes will make an automatic movement (called a saccade) in the direction of the stimulus. This is an example of an automatic response, and it requires effort to override it. It's a bit like a go/no-go task for eye movements. The ability to inhibit automatic eye movements develops throughout childhood and adolescence and stabilizes in early adulthood, and the prefrontal activity associated with this task also changes with age.

Since these first studies, many other experiments using versions of the go/no-go paradigm have confirmed that the ability to inhibit automatic responses is still developing in adolescence, as are the brain regions involved in this cognitive ability.

∞

Even though brain *development* seems to level off at some point, the brain never stops being capable of change. *Plasticity* – the brain's capacity to adapt to changing environmental stimuli – is in action all the time, whenever learning takes place, and there's no age limit on it. Plasticity enables us to learn, and we can learn new information at any age. So what exactly is the difference between brain development (which occurs throughout childhood and adolescence and into early adulthood) and brain plasticity (which operates at all ages)?

To understand this, we need to think about two different types of plasticity. *Experience-dependent plasticity*, as it's sometimes known, is the brain's ability to adapt to new information and underlies new learning at any age. In contrast, *experience-expectant plasticity* is the readiness of the brain to respond to sensory input from the environment during development. The latter type relates to the idea of the 'sensitive period'. An example of early experience-expectant plasticity at a sensitive period is the mechanism by which an infant learns the sounds of his or her language, as discussed in chapter 4. For some functions to develop normally, such as recognizing the sounds of one's own language, the baby must receive appropriate sensory input from its environment at the appropriate stage of development – the sensitive period. If a child is deprived of a particular kind of input (say, speech) during that important period, it will be difficult for that function (in this case, language sound perception) to develop normally.

During the sensitive period, synapses can be pruned or strengthened, but after the sensitive period, very little change to synapses can occur. Most sensitive periods that have been studied in early development pertain to sensory processing and motor development. During sensitive periods in early development, the brain 'expects' to be exposed to lots of different sensory input that is present in almost every environment, including visual stimuli, noises and speech sounds, and objects to be moved and manipulated. This is experience-expectant plasticity.

What about after early development? Perhaps this experience-expectant plasticity continues into adolescence for certain functions and certain brain regions. It seems reasonable to predict that there would be later sensitive periods for cognitive processes that rely on brain regions undergoing significant development during adolescence – processes such as reasoning, planning and social cognition.

A training study carried out in my lab showed that the learning of certain cognitive skills is better at certain points in adolescence than at others. This study was carried out by Lisa Knoll, Delia Fuhrmann and Ashok Sakhardande. It was a large study that took several years to complete, involving 558 adolescents aged 11–18 and 105 adults aged 19–33. Participants were divided at random into three training groups, and each group completed up to twenty days of online training, for ten minutes per day, in a single cognitive skill.

The training was designed to be fun: it involved computerized games, designed by a professional software company in collaboration with us, that could be played on a computer or tablet, at home or at school. The games become progressively harder as the player's performance improves, but no advice is given about *how* to improve – instead, learning relies on trial and error, feedback and practice. As you are playing, you're given feedback about how well you're doing – words like 'awesome' and 'fantastic' if you're doing well, or 'better luck next time' if you're not doing so well. At the end of each training session, you get to open a treasure chest and find out what treasure you have won – the aim being to collect as much as possible. In addition, we paid participants a small amount of money to motivate them to keep on training.

One training group practised non-verbal reasoning, which is the ability to detect relationships between objects and is related to intelligence and mathematics performance. The non-verbal reasoning task involved looking at a 3 × 3 grid of abstract shapes, which could vary by colour, size, shape and orientation. The final cell of the grid was left blank (see illustration opposite), and participants had to choose the correct shape to complete the pattern.

The second group were trained on a 'numerosity discrimination' task, in which participants were shown two groups of different coloured dots in quick succession on a computer screen and had to judge which group had the most dots. This is a skill known to be related to mathematical ability.

The third group of participants were trained on a face-recognition

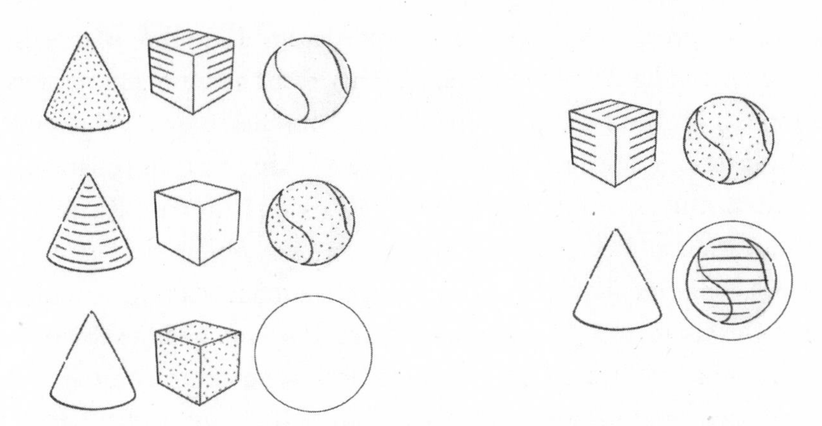

An example of the non-verbal reasoning task, in which participants are asked to identify the missing piece from the four on the right to complete the grid. The correct answer is circled for illustration

task, which was included as a control. We knew from previous research that face-recognition performance would be likely to change during adolescence, but we didn't think it would be particularly amenable to *training* at this age.

Before and after training, all participants were tested on the three cognitive skills: non-verbal reasoning, numerosity discrimination and face recognition. Participants were also tested six months later to see whether the effect of training had lasted.

We divided the participants into four age groups (early, mid- and late adolescents, and adults). Our results showed that all age groups improved their performance when trained in non-verbal reasoning, but late adolescents (aged 16–18) and adults (aged 18–30) showed the greatest training benefits. Training in the numerosity task also yielded some improvement in performance, but only in late adolescence and adulthood. Training in face recognition didn't have any effect on performance in either adolescents or adults, which is what we expected.

These findings suggest that training in non-verbal reasoning and numerosity, both of which are related to mathematics performance in schools, has greater effects in late adolescence (and adulthood) than in early or mid-adolescence. We made sure this wasn't because some groups did more training than others. The results from this study highlight the relevance of this late developmental stage for education and challenge the widespread assumption that 'earlier is always better' for learning.

Education policy tends to emphasize the importance of early childhood interventions. To quote the US National Scientific Council on the Developing Child from 2014: 'Brain plasticity and the ability to change behavior [learning] decreases over time.' This argument is partly based on the findings from economics that interventions early in life are more worthwhile than later interventions in terms of the money saved. However, this emphasis on early interventions is at odds with the findings that the human brain continues to develop throughout childhood and adolescence and into early adulthood.

Of course, early interventions are vitally important. But if children are to thrive they need nurturing throughout their childhood and adolescence. While nurturing environments, learning opportunities and emotional support in early childhood are vital, the key point is that this support should be continued throughout development, because learning doesn't end with childhood. If a child has 'slipped through the net' and hasn't done well during childhood, it's not too late to start intervening and providing extra support in adolescence.

The results of our training study highlight the importance of *late* adolescence for education, and the need to investigate late adolescence as a potential window of opportunity for educational interventions. They suggest it might be more efficient to wait until late adolescence to learn certain types of cognitive skill.

In our study, although it was adults and late adolescents who benefited *most* from training in non-verbal reasoning, the average test score for early adolescents aged 11–13 improved from 60 per cent to 70 per cent following three weeks of ten-minute daily online training sessions. This calls into question the claim made by some education policy-makers that entry tests for selective schools that include non-verbal reasoning assess the true potential of every child, which is fixed and possibly innate. On the contrary, this is a skill that can readily be trained and improved.

∞

So we know that the adolescent brain is still primed for change; in adolescence we are sensitive to experience and training. This isn't to say, of course, that once late adolescence is over, that's the end of our learning; we know from our own lives that isn't true. And indeed, the brain retains a certain level of plasticity throughout life. This is the *experience-dependent plasticity* I mentioned earlier, and it is different from *experience-expectant plasticity* because it is reactive to learning. Experience-dependent plasticity in the adult brain generally occurs as a function of usage. In other words, the adult brain continuously adapts to changes in its environment. Learning to play tennis or a musical instrument, learning new vocabulary or new computer software: all these are examples of tasks that require experience-dependent plasticity. There is no age limit to this kind of learning, which can and does occur throughout life.

In fact, the brain's experience-dependent plasticity is constantly in use, whenever a new memory is laid down or a new face is seen, for example – even as you read this book. In chapter 5, I described how experience-dependent plasticity is illustrated by the hippocampus of London cab drivers. Further examples come from music. The part of the brain that processes sound (the auditory cortex) in highly skilled musicians is about 25 per cent larger than it is in people who have never played an instrument. The degree of enlargement is correlated with the age at which musicians began to practise, suggesting that the expansion of the auditory cortex is dependent on how much it is used.

Brain changes can be rapid – even occurring over a matter of days. In a study of how the sensory and motor areas of the adult brain can adapt according to how they are used, adults who were new to the piano learned a simple piano exercise for two hours a day for five days. The area of the brain that controls finger movements (the motor cortex) increased in volume and became more active in these participants compared with a control group who had not learned the exercise.

Such changes may also be reversible. A number of studies have looked at learning to juggle. One of these studies, carried out by Arne May and colleagues at the University of Regensburg in Germany, scanned people's brains before and after they had practised juggling three balls every day for three months. At the end of this time, two regions of the jugglers' brains that process visual motion information had increased in size. But after the passage of another three months, during which the same people had not done any juggling, these regions had returned to their previous size.

Experience-dependent plasticity, then, continues throughout life and comes into effect whenever you learn something new. Let's try this now. You might not know that the distress call 'Mayday!' comes from the French '*M'aidez!*,' meaning 'Help me!'; or that the English word 'porcupine' comes from the French *porc épine* ('spiky pig').* Try to learn these – if you succeed, a few thousand synapses have been strengthened in your brain, and this might enable you to remember these facts for a long time (although you will probably need to revise your new knowledge).

This and other similar abilities are all individual skills. Equally important for healthy development are the social and communicative skills we have already touched on when considering the developing sense of self in adolescence. In the next two chapters, I'll turn to look in more depth at what we know about the development of these complex attributes in these years when one's place among family, friends and – especially – one's peer group comes to take on huge importance.

* For these and many more examples, see the video at: http://mentalfloss.com/article/74782/french-phrases-hidden-english-words.

7

Social mind, social brain

HUMANS ARE INTENSELY SOCIAL. We recognize faces from the moment we are born, we seek out social interactions and gradually build an understanding of other people's minds, we learn language seemingly effortlessly in our first few years, and our social experiences become richer and more complicated as we progress through childhood and adolescence. Imagine what life would be like without access to social experiences. So essential to our happiness and well-being is the freedom to mix with other people and form meaningful relationships that to take that freedom away, through imprisonment or, in the most extreme form, solitary confinement, can be viewed as the ultimate punishment short of death.

Given that our social world is quintessentially important to us, it is perhaps not surprising that a large proportion of the human brain is dedicated to understanding other people and to social interaction. Social understanding involves a complex set of cognitive processes, which we can call the *social mind*; and the social mind in turn relies on a network of brain regions referred to as the *social brain*. The social brain enables us to recognize others, evaluate their intentions and feelings, and predict their future actions. Some social brain regions are involved with perceptual processes such as the recognition of faces, voices, bodies and movement – walking, running and dancing. Other social brain regions are involved in more complex processes such as recognizing other people's emotions and mental states, and understanding another person's perspective.

∞

One morning you are standing on a station platform waiting for a train. You see a man in a suit running quickly towards the next platform, where the train is just leaving. The man waves his arms in the air and shouts, 'Please wait!' but it's too late: the train has gone. A look of panic crosses his face. A few minutes later, his mobile phone rings. He takes it out of his pocket and looks at the display screen. His face goes red and he looks even more panicked; then he answers it and says, 'Oh, hello, Mr Jacobs. I'm terribly sorry but I won't be able to make the meeting because my train has been cancelled. Typical British trains!'

You might conclude from watching this scenario that the man needed to get to a meeting, is annoyed that he missed his train and becomes worried when someone (his boss, perhaps?) phones to check where he is. You know he is lying about the train being cancelled.

In this complicated situation, your *social mind* has – rapidly and instinctively – employed many different social cognitive processes to interpret the scene you've witnessed. First, you've recognized the person as another human. This might sound trivial, but the recognition of faces and bodies as belonging to members of the same species entails a complex set of computations by the brain. Second, you've recognized emotional expressions on the man's face and in his voice. Humans are capable of distinguishing a surprisingly large array of emotional expressions, from basic emotions such as joy and fear to complex emotions such as deviousness, jealousy and shame. Third, you've made inferences about what is going on in this person's mind. We are constantly reading other people's faces and actions to try to work out what they are thinking and feeling. This ability to attribute thoughts and feelings to other people is known as *theory of mind*, or *mentalizing*.

Most people mentalize automatically and effortlessly whenever they see or interact with another person. Without the ability to do this, other people would be difficult to decipher. This is especially the case for people on the autism spectrum. Autism is characterized by difficulties with social interaction. At the 'low end' of the spectrum, the ability to communicate is severely affected and general cognitive functioning is low. The 'high end' of the spectrum is characterized by difficulties in social interaction together with average to high intelligence. Other people – especially what they are thinking and feeling – are often an enigma to people on the autism spectrum. To explain these difficulties, the *mind-blindness* hypothesis of autism was proposed by Simon Baron-Cohen, Alan Leslie and Uta Frith at the University of London in the 1980s and has been developed since by many other researchers. The main proposal of the mind-blindness theory is that the intuitive understanding that other people have minds is lacking or diminished in people with autism. If people with autism cannot automatically mentalize, then this would explain why they find communication and social interaction, especially understanding the nuances of social interaction, so challenging. Some people with autism seem to be unable to understand that other people can have different beliefs and intentions – different mental states – from their own, and this inability becomes apparent early in life. We'll return to this point later in the chapter.

Typically developing children acquire an understanding of the social world very rapidly; indeed, a basic understanding of other people is present at birth. In order to study the perceptual and cognitive abilities of newborn babies, there are science labs in maternity wards, aimed at capturing information from those first few hours of life and learning about early brain development. Studies carried out in this setting have shown that babies are able to distinguish between different visual objects from the very first day of life. We know this because scientists have found that newborn babies gaze for longer at a new object than at an object which is shown to them multiple times, which causes them to become bored and look away. They also look longer at things they prefer. This forms the basis of the *looking test*, which has been used in many studies of perception and cognition in babies. It is an important and useful test, because it provides a window into baby cognition that doesn't rely on verbal interaction – obviously, at this stage of development, babies aren't able to tell us explicitly what they're perceiving or thinking.

∞

The maternity ward studies have also shown that babies are born with a rudimentary – but remarkable – capacity to recognize faces. One study found that newborn babies look for longer at drawings of whole faces than at drawings of faces that have been 'scrambled' – in which all the features are in the wrong places – even though their exposure to faces has, at most, been only very limited. This suggests that, at birth, the brain might be pre-programmed to recognize faces as being different from other objects. And within just a few days, a baby learns to recognize his or her mother's face, staring at photos of her for longer than at photos of other people's faces. John Morton and Mark Johnson, developmental psychologists at UCL and Birkbeck in London, have studied face recognition in babies for many years. They were interested in identifying which brain regions are responsible for humans' very early ability to recognize faces, and suggested that the brain pathways involved in face recognition during the first few weeks of life differ from those employed in later face processing. Early face recognition has been found to rely on subcortical structures. These structures, most of which develop early in life, contribute to a pathway in the brain that enables us to make fast, automatic movements in response to what we see or hear. The same subcortical structures are present in many other animals, enabling them to recognize and respond quickly to potential threats such as predators, as well as to identify figures of safety such as parents, and to spot prey.

Johnson and Morton suggested that our early ability to recognize faces may have evolved because it allows newborn babies to develop an automatic attachment (this is called 'imprinting') to the people with whom they spend the most time, and who provide them with food and protection. Just as a newly hatched chick imprints on its mother and follows her around wherever she goes, newborn human babies imprint on the face they see most. It is only when babies reach 2 or 3 months that *cortical* brain regions, on the surface of the brain – mostly in the temporal and occipital lobes – start to take over the control of face recognition.

As well as our ability to see, our brain's hearing (auditory) system is also already partially developed at birth. In fact, research with premature babies, born during what should have been the third trimester of pregnancy, has shown that even at this very early age babies are already responsive to the sound of human speech and will listen longer to this than to other sounds.

We know about babies' hearing abilities from studies carried out in maternity wards in which scientists measure how much the baby sucks on a pacifier. The idea is that the baby will produce longer sucks for sounds that are not familiar, and shorter sucks for sounds he or she recognizes. Using this clever method, scientists have discovered that, by day one of life, babies can distinguish between male and female voices. By day two, they can detect the difference between a foreign language and their own – sucking longer when they're exposed to foreign-language sounds than when they hear sounds from their native language, to which they were exposed in the womb. By day three, a baby can recognize his or her mother's voice, preferring to listen to her speech than to that of strangers. Such remarkably early hearing skills are thought to emerge from the ability to hear some – albeit muted – sounds in the womb, when the developing brain is rapidly building connections and learning how to interpret sounds. This ability forms the foundation for the building blocks of language, which is a fundamental component of social interaction.

The majority of research on the development of face processing and sound recognition has focused on early childhood. Is there any further development of social perception in adolescence? We might expect there to be changes in the social mind as the social world becomes vastly more complicated, and the importance of friends, peer groups and peer influence increases, during and after puberty.

Experiments in the 1980s, led by Susan Carey and Adele Diamond at Harvard University, showed that face processing continues to change in early adolescence, and in intriguing ways. Male and female participants aged 8–16 were asked to remember the identity of unfamiliar faces from photographs. Performance on this face memory task improved rapidly between 8 and 10 years, levelled off for a couple of years, and then actually *declined* at around age 12, followed by gradual improvement, back to pre-age-12 levels, up to age 16. What was going on? It seems counter-intuitive that performance on this task should become *worse* with age in early adolescence.

In a second study, Carey and Diamond discovered that the dip in face memory was not driven by age alone. They asked a large group of 10- and 11-year-old girls to carry out the same face memory task as before, and they also measured what stage of puberty the girls had reached. At this age, some girls will be in the early stages of puberty, others in mid- or even late puberty – this degree of variation is completely normal and expected within this age range. The researchers found that girls in mid-puberty performed less well on the face memory task than girls of the same age in early puberty. This suggests that something happens around mid-puberty that temporarily disrupts face processing.

Another behavioural study carried out in the 2000s by Robert McGivern and his team at San Diego State University reported similar results using a rather different task in an experiment carried out with a large group of children, adolescents and young adults. In this task, participants looked at pictures of faces showing particular emotional expressions (happy, sad, angry), or words describing those emotions ('Happy', 'Sad', 'Angry'). They were asked to identify, by pressing a button corresponding to each emotion as quickly as possible, the emotion presented in the face or word. Following this, the participants were shown a face and a word at the same time, and had to decide whether the facial expression *matched* the emotional word. This second task – called a *match-to-sample* task – involves the prefrontal cortex, and is harder: participants are generally slower to match a word to a face than to respond just to a word or to a face.

When the researchers analysed how long it took the participants to answer the questions, they uncovered an intriguing result. At the age of puberty onset, around 11–12 years, there was a *decline* in performance in the task in which participants had to match faces and words, compared with the younger children aged 10–11: the 11–12-year-olds were about 15 per cent *slower* at performing this match-to-sample task than the younger group. The results add weight to the suggestion that there is a dip in performance on tasks that involve categorizing emotional faces at the onset of puberty. After puberty, from age 13–14 years, performance improved until it returned to the higher, pre-pubescent level by the age of about 16 years.

We don't yet understand what causes the possible dip in performance at puberty. It's possible that the large changes in sex hormones at this time might trigger changes in brain circuitry, and there is some evidence to support this hypothesis. Research carried out by Anne-Lise Goddings, Kate Mills and our collaborators from NIMH has shown that the development of subcortical brain regions progresses in line with the course of puberty rather than with chronological age. What this means is that subcortical development is associated with how physically mature you are, rather than just how long you have been alive.

The researchers who carried out the match-to-sample study linked the pubertal dip in performance with the sudden increase in the number of synapses – connections – between neurons in the brain at the beginning of puberty. You will remember from previous chapters that both developmental MRI and post-mortem brain studies have shown that the structure of the prefrontal cortex varies at different stages of life. One of these phases of change occurs in late childhood and early adolescence, when a rearrangement of nerve cells and their connections takes place. The number of synapses in the prefrontal cortex increases steadily throughout childhood, reaching its highest point at some point in late childhood; after this point, the pattern reverses and the number of synapses declines, with synaptic pruning eliminating connections that are no longer required, and strengthening those that continue to be used. Some scientists have proposed that the new synapses created during childhood are not yet specialized for processing certain stimuli, so that many will eventually be pruned away as unnecessary. It is possible that, during late childhood and puberty, the presence of excess synapses results in a temporary decline in performance while the brain reorganizes. Only later are these extra connections pruned into efficient networks, organized to perform all the specialized functions required. Puberty occurs around eighteen months later in boys than in girls, on average, which might explain why the dip in performance in the McGivern study happens a little later in boys than in girls.

It is also possible that, from puberty onwards into early adolescence, the social world around us – other people, what they look and act like, how they treat us, our place in the social group – also influences the pruning process, helping to determine which connections are retained and which are deemed unnecessary. If this is so, the social environment plays a role in shaping the developing brain. However, this is a very speculative suggestion, and we don't yet know much at all about what causes the reorganization of synapses in some brain regions at puberty, or how this affects perception and cognition.

Developmental dips don't just occur for face processing. In a study carried out by Iroise Dumontheil, who was working with me at the time and now runs her own lab at Birkbeck in London, we detected a dip in performance in early adolescence in a non-verbal relational reasoning task. This task – called the Shapes Task – involves assessing whether two pairs of objects, which can vary in shape and/or texture, change along the same dimension or along different dimensions. Look at the illustration below. Do the items in both pairs differ in texture (one dimension), or in shape (the other dimension)? Or do they differ in different dimensions? This is a difficult task that relies on the rostro-lateral prefrontal cortex (RLPFC) and the parietal cortex. We tested 179 female participants aged 7–27 years and found a non-linear pattern of development across age. After an early improvement in accuracy, with 9–11-year-olds performing at adult levels, performance dipped in the 11–14-year-olds and then gradually improved again to adult levels throughout late adolescence. This pattern of performance was mirrored by a non-linear development in brain activity in the left RLPFC on the same task.

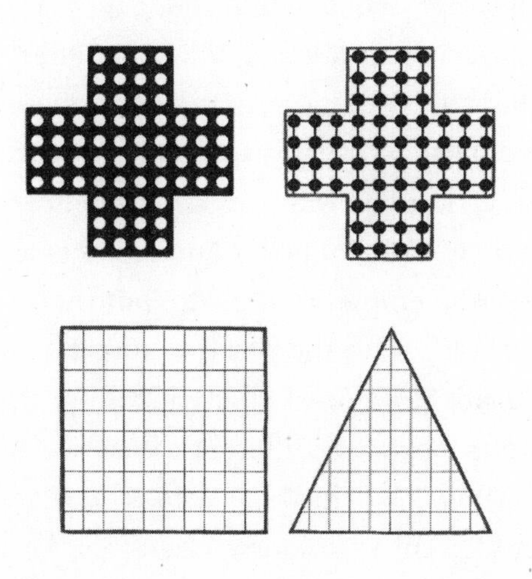

An example of the stimuli in the Shapes Task: the items in the top pair differ on the texture dimension, while the items in the bottom pair differ on the shape dimension

It's still not clear what implications this early-adolescent performance dip in certain reasoning and perceptual tasks could have in everyday life. However, there is evidence for a similar phenomenon in educational performance – referred to by teachers and education researchers as an *educational dip* in early adolescence. At this time – somewhere between 12 and 14 years – some students do worse at school than they did in the previous year. Then they gradually improve again in the year that follows. This educational dip is intriguing and puzzling, and educational researchers have suggested it might be a consequence of the pressure of moving from small junior schools to large senior schools, which can be stressful socially and academically. But perhaps it's more than that. Perhaps the educational dip, as well as the dip in face processing and reasoning found in the studies described above, is in part a reflection of changes in the brain that occur in early adolescence.

∞

At the beginning of this chapter I introduced the concept of mentalizing, the ability to understand other people's mental states. How does mentalizing develop? Within the first nine months of life, babies develop a basic understanding of other people's emotions. They are able to tell the difference between facial expressions reflecting emotions such as happiness, sadness and anger, and to recognize that these are associated with different feelings. However, until around 18 months, babies act in a seemingly egocentric way – as if they believe that everybody shares the same viewpoint, thoughts and feelings as themselves. Babies do not yet have the empathic skills to 'put themselves in someone else's shoes', to identify with the way that somebody else is feeling. Their brains have not developed that capability. At around 18 months, however, babies begin to differentiate between their own feelings and those of others. By this age, babies start to realize that different people have different emotions and desires. The ability to understand these concepts continues to become more sophisticated through the first few years of life and beyond.

In order to determine at what age babies develop the concept that other people have different opinions from their own, Alison Gopnik from the University of California at Berkeley devised some clever experiments. In one test, babies aged 14–18 months were shown two different types of food: broccoli and cookies. Perhaps unsurprisingly, babies usually chose to eat cookies rather than broccoli when given the choice in the first part of the experiment. The experimenter then ate the broccoli in front of the babies and looked as if she really enjoyed it, and then ate the cookies while looking disgusted, as if she didn't enjoy them at all. When they were subsequently asked to pass her some food, the 14-month-old babies still passed the experimenter the cookies – the foods they preferred – despite her demonstration that her preference was different from their own. In contrast, by 18 months, babies passed her the broccoli – showing that by this age, babies have started to realize that other people's tastes may differ from their own. Their view of the world is becoming less egocentric.

As soon as they are able to speak, children begin to talk about their minds and those of others. At first, they tend to concentrate on their desires, perceptions and emotions: what they see, feel and want takes precedence over what they think or know. By 2 years old, children begin to communicate both verbally and non-verbally with other people to indicate their desires. They might point at the cookie tin and say, 'I want a cookie.' By this age, children also start to understand that the direction in which you gaze can provide information about what you want. They'll tell you that their sister is staring at the cookie tin because she would like a cookie, even if she hasn't told them so.

By 3 years old, children start to talk about their beliefs: 'I *think* the cookies are in the jar.' However, it is not until about 4 or 5 years that they are able to verbalize the beliefs of others, and to acknowledge that these might be different from their own: 'My sister *thinks* the cookies are still in the jar but I *know* they're not.' This ability to understand that people's beliefs differ from one's own beliefs, and can be false, is a component of mentalizing – our ability to form views about other people's minds.

Developmental psychologists have designed theory of mind tasks that involve demonstrating an understanding that other people can have a false belief. The most famous false-belief task, devised in the 1980s by Simon Baron-Cohen, Alan Leslie and Uta Frith, is the *Sally–Anne Task*. In this task, the child is shown two dolls, Sally and Anne, in a room containing a basket and a box. The researcher shows the child that Sally has a marble, which she puts in the basket. Then Sally leaves the room. While Sally is out of the room, Anne removes the marble from the basket and hides it in the box. The child is asked where Sally will look for the marble when she returns. Three-year-olds don't do well on this task – they almost always say that Sally will look in the box, where *they* know the marble is hidden. In other words, at age 3, children don't seem to understand that another person can have a false belief, which differs from their own belief and from reality. By 4 or 5 years, children give the correct answer (the basket), suggesting that by this age they have developed a theory of mind and understand that people can have false beliefs.

Children with developmental conditions such as autism often have difficulties understanding other people's beliefs, and many fail the Sally–Anne test at 5 years or older. As I mentioned earlier, Uta Frith and her colleagues have suggested that it is this difficulty in understanding other people's minds which leads to the problems in forming social relationships in children with autism.

For several decades from the 1980s, it was generally agreed that theory of mind in typically developing individuals appears between 3 and 5 years of age. However, this consensus was challenged in the 2000s, when new research demonstrated that false-belief understanding is present in much younger children. Renée Baillargeon and her colleagues at the University of Illinois showed that babies as young as 15 months could be aware of another person's false belief. To do this, they had to find ways of testing infants appropriate to their age. At 15 months, infants can't explain verbally to the researcher where another person is likely to look for the hidden object. But by such means as recording eye movements, it is possible to glean an idea of what babies are thinking.

Baillargeon and Kristine Onishi measured the eye movements of babies when they were interacting with a researcher. In one study, babies played with a researcher who, at a certain point in the game, hides a toy. When the researcher isn't looking, the toy is moved by another adult and hidden in a different location, so the researcher has a false belief about where the object is. Where will the researcher look for the toy? The baby is too young to tell us, but we can get an idea about where they expect the researcher to look for the object by measuring their eye movements. If the researcher looks in the toy's actual location (even though she can't know it's there), the baby might be surprised and look longer at this scene. We can also guess the baby's understanding of the researcher's mind if the baby looks before the researcher does in the location where the researcher should search based on her own belief about where the toy is (and not where it actually is). Again, in this case of anticipated looking, the baby may be demonstrating an understanding of the researcher's false belief. This is exactly what Baillargeon and her colleagues found: by 15 months, babies looked towards the place where the researcher should search for the object, even though the object wasn't actually there. Babies of this age, then, seem to have some implicit understanding of the contents of another person's mind.

Another study, carried out at Birkbeck in London by Atsushi Senju, Victoria Southgate and their colleagues, monitored the gaze of 18-month-old babies. This confirmed that infants of this age can anticipate, as demonstrated by the direction of their gaze, that a researcher will approach the location where she saw a toy being hidden, even if the baby has seen that the toy has been moved and is no longer in the same place. At around this age, babies are also able to respond appropriately to prompts to help the experimenter, such as 'Go on, help her!' Following such instructions, they often show the experimenter where the toy is hidden by pointing.

A further series of studies by Michael Tomasello and his colleagues in Leipzig, Germany, demonstrated wonderfully how 18-month-old toddlers are able to read the hidden mental states of adults. These researchers invited parents to bring their 18-month-olds to the lab, and videoed them while they were in the waiting area. While the parent and child were waiting, a man they had not met before (one of the experimenters) entered the room with his hands full of heavy books. He walked towards a cupboard and made an obvious attempt to put the books in the cupboard, but was unable to open the doors of the cupboard because his hands were full. He pushed the books into the cupboard doors, and looked slightly frustrated, but didn't say anything. More often than not, the toddler would spontaneously walk over to the cupboard, open the door and then make eye contact with the man, as if to say, 'There you go, you can put the books in now.' The researchers reasoned that the toddlers must have understood the intentions of the man – to open the cupboard and put the books inside – and it was this ability to mentalize that elicited their spontaneous helping behaviour.*

∞

So, decades of developmental psychology research have shown that an understanding of other people's minds develops gradually over the first five years of life. Until the mid-2000s, there was not a lot of evidence that

* There are some wonderful videos of the toddlers (and also chimpanzees!) being helpful at: http://www.eva.mpg.de/psycho/publications-and-videos/study-videos.html.

the ability to mentalize changes much after mid-childhood, around 5 or 6 years. But this notion was increasingly at odds with what was rapidly being revealed about the development of the social brain. New MRI studies (see chapter 8) were showing that many brain structures, including those responsible for mentalizing, undergo substantial development well beyond early childhood, right through adolescence and into early adulthood. If the brain's structure changes in adolescence, how does this affect the ability to mentalize? This was a question I began to investigate in the mid-2000s.

One difficulty soon presented itself. Most experimental tests of mentalizing had been designed to investigate this ability in young children, and result in near-perfect performance if administered after mid-childhood. This is a problem if you want to study a cognitive ability after that age, because a task that is too easy, and on which everyone performs perfectly, cannot reveal the more subtle changes in this ability with age. What we needed was a task that involved mentalizing but was challenging, even for adults, and could therefore show variation in performance across development.

Our answer came in the form of a clever experiment called the *Director Task*, which was designed by Boaz Keysar at the University of Chicago. The task requires the participant to take into account the mental state (in this case, the visual perspective) of someone else when communicating with them. The original version of this task involved the participant communicating with a real person (called 'the Director'), who instructed the participant to move objects that were located in a set of shelves. The participant stood in front of the shelves and could see all the objects. The Director stood behind the shelves, and some objects were hidden from her point of view by a piece of wood behind some of the shelves. Therefore, the Director could see some but not all of the objects that the participant could see. Participants were told that the Director would not ask them to move objects she couldn't see.

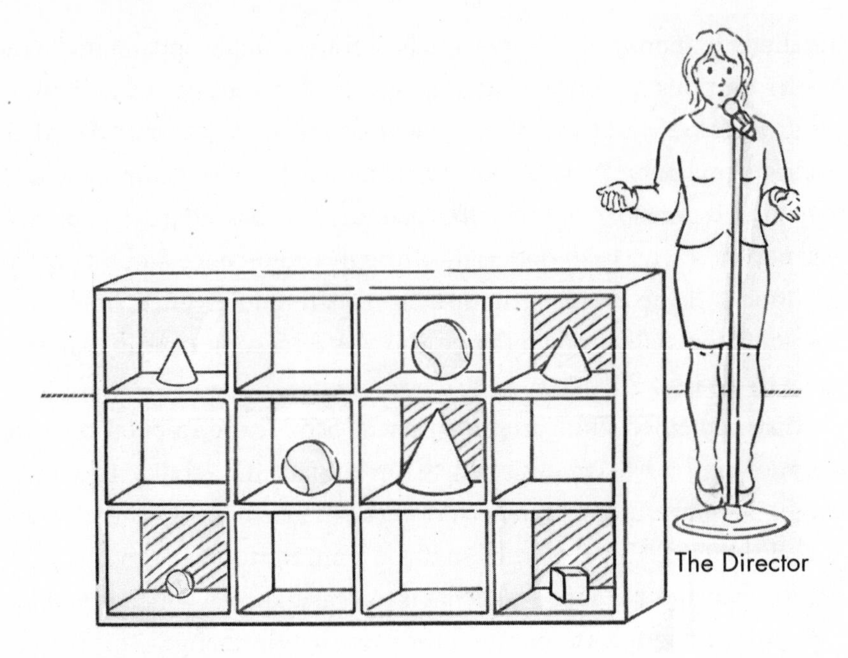

The Director Task

During the task, sometimes the Director asked the participant to move an object they could both see, and sometimes she asked the participant to move an object of which there were three examples, two she could see and one she couldn't. In the example shown in the illustration on this page, the Director might say, for instance: 'Move the small ball up one shelf.' In this example, there are three balls: the smallest isn't visible to the Director, so the ball she must intend the participant to move is the next size up, which is visible to both the Director and the participant. So on this trial, the correct ball to move would be the medium-sized ball.

In Keysar's study, this task resulted in a surprisingly large number of errors – in about 50 per cent of trials, adults made mistakes and moved an object that the Director couldn't actually see (in the example above, the smallest of the three balls). This suggests that taking into account someone else's perspective when making decisions is challenging, and that our own perspective easily interferes with our ability to take into consideration someone else's.

Iroise Dumontheil and I, together with Ian Apperly at the University of Birmingham, used a computerized version of the Director Task designed by Ian. In this version, just as in the original version, the participant views a set of shelves containing objects, and is instructed to move objects by a 'Director' standing behind the shelves, who can see some but not all of the objects the participant can see. Again, just as in the original version, sometimes the Director's request required participants to inhibit their egocentric response (when the critical object was hidden from the Director), and sometimes it didn't (when the critical object was visible to both the Director and the participant).

We also designed a second version of the task, in which the shelves looked the same but the Director was no longer there, and participants had to follow a simple rule: never move objects that have a dark grey background. This is effectively the same as the Director version, except that, when deciding which object to move, participants needed to remember the rule, rather than taking into account the Director's perspective. In other words, it's a non-social version of the Director Task.

We tested 177 participants aged 7–27 years on both versions of the task, and found that performance in both tasks improved gradually between late childhood and mid-adolescence. This means that the processes involved in both tasks – remembering and following instructions, inhibiting the tendency to move the first object that comes to mind (e.g. the smallest of the three balls) and integrating the auditory instructions with the visual perception of the shelves – all improve gradually up to mid-adolescence. What happened after mid-adolescence, though, was intriguing. While there was no further improvement with age on the non-social task (remembering the rule about the grey background), accuracy in the Director Task continued to improve between late adolescence and adulthood. These results suggest that the ability to take another person's visual perspective into account when making decisions and actions is still undergoing development at this relatively late stage.

That said, even adults in our study didn't do well on the Director Task, making errors in around 45 per cent of attempts at the social task. This was despite the fact that they were intelligent people – the average IQ of our adult participants was around 119, well above the average IQ of the population, which is 100. It seems that taking other people's perspective into account doesn't always come easily, even to intelligent adults.

∞

There are other aspects of social cognition that change during adolescence, including how much we trust other people and cooperate with them, and how much we seek out fairness. These aspects of social decision-making have been studied in *behavioural economics* games, which involve making decisions about money.

In the *Ultimatum Game*, there are two players: these can be two real participants in the laboratory, or one participant in the laboratory and another (real or fake) participant over the internet. Player One (the participant) is given an amount of money by the experimenter, let's say £10, and is told he can keep it all for himself or split it with Player Two. Although there are huge individual differences in how much people donate, most players do give some of their money to the other player, even if they know they'll never see the other player again, or indeed meet them at all. Critically, once Player One has decided how to split the money with Player Two, Player Two can decide whether to accept or reject Player One's offer. If the offer is accepted, the money is split accordingly, but if it is rejected, neither player receives anything. Interestingly, sometimes Player Two rejects the offer if they consider it too low, meaning they get less than they would have done had they accepted the low offer. This seems irrational, and is presumably motivated by a sense of fairness. The Ultimatum Game is interesting because it requires Player One to put themselves in the shoes of Player Two, consider what Player Two will think of the offer they make and consider what they might do as a consequence – in other words, it involves mentalizing.

A study carried out by Berna Güroğlu and Eveline Crone and their colleagues at the University of Leiden in the Netherlands used a version of the Ultimatum Game to investigate changes in behaviour relating to fairness during adolescence. The researchers wanted to find out whether adolescents make fairer decisions than children because of an increased understanding of other people's intentions and a greater consideration of what would constitute a 'fair' decision – both for themselves and for the other player.

Each participant played several rounds of a game in which a computer generated financial offers from a character called the Proposer. The participants thought the Proposer was another participant playing the same game in a different room – this was important, because the researchers wanted participants to feel they were making real offers to real people. In fact, this was just a cover story. Programmed by the experimenter, the Proposer (in the role of Player One) put forward a variety of offers to the participant: some offers were generous and some were not.

Participants aged 9–18 years were told they could choose to accept or reject each offer (in the role of Player Two). The researchers didn't mention anything about the need to consider what was fair – participants were just told the rules of the game, and specifically that if they accepted an offer, the money would be shared as proposed, whereas if they rejected an offer, neither player would receive anything. The participants were shown what choice the Proposer had: sometimes they were told that the experimenter had not allowed the Proposer to do anything but make an unfair (low) offer. At other times, participants were led to believe that the Proposer had a choice between making a fair and an unfair offer.

The results demonstrated that adolescents aged 12–18 were more likely than younger children to take into account the Proposer's intentions (and the restrictions placed on the Proposer by the experimenter) when they decided whether or not to accept an offer. In other words, they were more likely than younger children to reject an unfair offer when the Proposer could have made a fair offer than when the Proposer had no choice. This suggests that the ability to integrate the perspectives and intentions of other people into one's assessment of fairness gradually increases during adolescence.

∞

The *Trust Game* is slightly different: it places a greater focus upon how your decision will affect another person, as opposed to how their decision will affect you. It requires the participant to consider how much they trust the other player, as well as what the other player might be thinking. Player One (the participant) is given an amount of money by the experimenter and is told he can keep it all for himself or split it with Player Two. The amount (if any) given to Player Two is then typically multiplied by the experimenter, so that Player Two might be given three times what was given to her by Player One. Player Two then chooses either to send some of the money back to Player One (this is called *reciprocating*), or to keep most – or all – of it for herself (*defecting*). If Player One doesn't trust Player Two, he might not share anything with her; if he does trust her, he's likely to give her some of the initial donation and hope that she will reciprocate so that he will end up with more money than his initial amount.

Anne-Kathrin Fett at the University of Amsterdam and her colleagues were interested in the relationship between perspective-taking and trust in adolescence. In her study, participants played both the Director Task (described above) and the Trust Game. Fifty adolescents aged 13–18 were divided into two groups according to how they performed on the Director Task. Half were classified as 'low perspective-takers', having been fairly poor at using the Director's perspective to decide which object to move. The other half were 'high perspective-takers', having been more adept at this task. The researchers found that the high perspective-takers demonstrated greater trust towards others, as evidenced by their initial investment in another player in the Trust Game, implying that those who are better able to consider another person's thoughts and actions are initially more trusting. While all adolescents modified their behaviour in response to unfair offers by decreasing investments and being less likely to share their money, high perspective-takers did so more drastically, suggesting a greater decrease in trust towards people who didn't play fairly. This suggests that adolescents who understood more about the other player's perspective and intentions were more offended by their subsequent low offer, and reacted accordingly. The results of this study indicate that increases in perspective-taking tendencies in adolescence may be associated with more sophistication and discrimination in deciding what levels of trust and reciprocity to adopt, as opposed to simply generalized increases in 'prosocial' (generous) behaviour.

Findings from Wouter van den Bos, Eveline Crone and their colleagues add support to this conclusion. They ran a series of Trust Games that involved dividing a sum of money, in the course of which they changed the identity of the partner with whom each participant was asked to share their money. The partner was either one of the participant's friends, someone they said they didn't like, a neutral classmate or an anonymous peer. During this experiment, younger adolescents (9–12 years) showed similar levels of prosocial behaviour (generous sharing) to all interaction partners, whereas late adolescents (15–18 years) showed increasing differentiation in prosocial behaviour according to their relationship with the partner, displaying the most prosocial behaviour towards their friends. This suggests that with age, *who* you are interacting with becomes more important. During later adolescence we start to place more weight on the identities of other people, perhaps because self-identity and how others view us become increasingly important to us as we establish ourselves as a member of our peer group.

The studies outlined in this chapter offer some fascinating discoveries and tantalizing suggestions about how the social mind develops after childhood and through adolescence towards adulthood. So what is going on in the brain as our perceptions and skills improve – and, sometimes, dip – during these years?

8

Understanding other people

THE FIRST BRAIN-SCANNING STUDY of mentalizing was carried out in the mid-1990s by Paul Fletcher, Chris Frith and their colleagues at UCL. In this study, adult participants had a brain scan while they read various different stories on a screen they could see while inside the scanner. Some of the stories – mentalizing stories – required an understanding of the story characters' mental states. Here's an example:

> A burglar who has just robbed a shop is making his getaway. As he is running home, a policeman on his beat sees him drop his glove. He doesn't know the man is a burglar, he just wants to tell him he dropped his glove. But when the policeman shouts out to the burglar, 'Hey, you! Stop!', the burglar turns round, see the policeman and gives himself up. He puts his hands up and admits that he did the break-in at the local shop.

Participants read the story and were then asked: 'Why did the burglar give himself up?' If the participant answered that the burglar must have assumed the policeman was on to him, then they were correctly using mentalizing to understand the situation. The task involves thinking about what the burglar knows, and what he doesn't know.

At other times during the brain scan, participants read a paragraph of unlinked sentences, and were asked a question about one of the sentences that appeared in the paragraph – this was the control condition. For example, they might read:

> The four brothers stood aside to make room for their sister, Stella. Jill repeated the experiment, several times. The name of the airport had changed. Louise uncorked a little bottle of oil. The two children had to abandon their daily walk. She took a suite in a grand hotel. It was already twenty years since the operation.

The researchers were then able to compare brain activity when participants were reading the mentalizing stories with brain activity when they were reading the unlinked sentences. Any brain functions that were involved in reading the stories *and* in reading unlinked sentences – such as seeing visual information, reading words, making sense of sentences and so on – would be eliminated in the comparison. Only brain regions that were *more* active when participants were reading the mentalizing stories than when they were reading the unlinked sentences would show up. In this way, the analysis revealed the brain regions that are specifically involved in mentalizing.

Fletcher and his colleagues found that four brain regions were more active when participants read the mentalizing stories rather than the unlinked sentences. These four regions are the dorsomedial prefrontal cortex (dmPFC), the temporo-parietal junction (TPJ), the posterior superior temporal sulcus (pSTS) and the anterior temporal cortex (ATC). This study was the first demonstration of the mentalizing network, and it paved the way for many further studies using different types of mentalizing task.

Fletcher's study used positron emission tomography (PET) scanning, which measures blood flow in the brain. When a group of neurons becomes activated, they require an increased amount of blood to replenish their supply of oxygen and glucose, which they need for energy. This regular supply of energy is crucial – remember, the brain takes up one-fifth of all the energy used by the body. The reliance of active neurons on glucose and oxygen transported to the brain in the blood is the principle underlying both PET and fMRI brain scanning.

PET involves injecting very small amounts of a radioactive chemical (called a 'tracer') into the participant's bloodstream, which carries the tracer around the body and into the brain. The tracer can be followed while it flows in the blood around the brain because it emits electrically charged particles called positrons. In a PET scan, the participant's head is surrounded by a camera that detects positrons and measures the location in the brain of the positrons emitted from the tracer. This indicates where in the brain neuronal activity is occurring. While PET was used quite a lot by researchers to study the human brain in the 1980s and 1990s, since the late 1990s it has largely been replaced by other brain-imaging techniques, in particular fMRI, that do not involve injections of tracers.

Since the PET study by Fletcher and his colleagues, a large number of fMRI experiments of mentalizing have been carried out by labs all over the world, using a broad variety of mentalizing tasks and stimuli. For example, Helen Gallagher and her colleagues at UCL carried out an fMRI study in which participants looked at cartoons. Some cartoons involved mentalizing – understanding the cartoon character's mental states – while others did not. When participants were looking at the mentalizing cartoons, the four regions of the mentalizing network were activated; this did not happen when they were looking at the non-mentalizing cartoons.

The consistency of results between studies is remarkable. Whenever participants mentalize, whether they are thinking about someone's mental states in a story, a cartoon, or a movie, the same four mentalizing regions become active. Even reading simple sentences that involve just a small degree of mentalizing is sufficient to activate the mentalizing network.

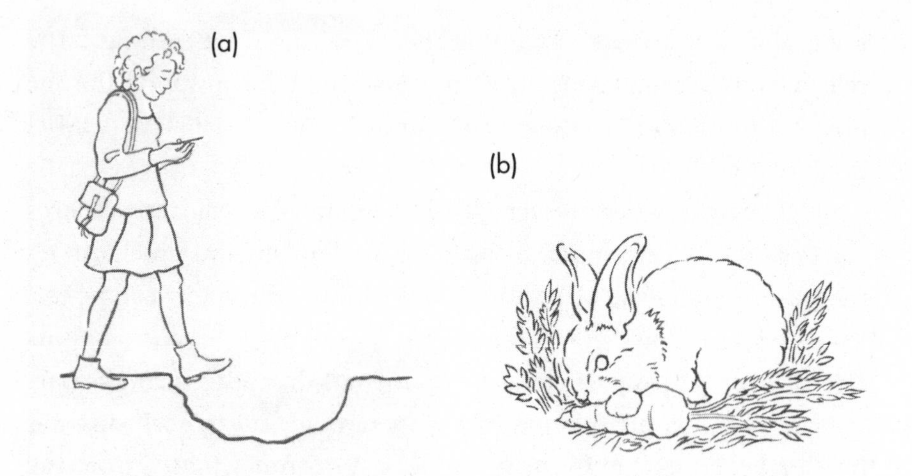

Examples of cartoons (a) involving mentalizing – will the girl reading her tablet fall into the hole because she's not paying attention to the ground? – and (b) not involving mentalizing – no understanding of the rabbit's mental state is called for

The consistency of results across brain-imaging studies of mentalizing is remarkable because mentalizing is a complex and rather vague concept: it's not like a more straightforward perceptual process such as recognizing simple visual shapes. It's perhaps less surprising that everyone's brain uses the same visual region for processing simple visual stimuli, but mentalizing involves many different component processes; it's a highly complex psychological skill.

What happens to the mentalizing network during development? In 2011, Kate Mills and I set up a study to find out. We were fortunate to be able to collaborate with Jay Giedd at the NIMH in Bethesda. Kate analysed brain scans from the 288 individuals in the NIMH study I discussed in chapter 6: all of them were aged between 7 and 30 years, and had undergone between two and seven brain scans, each around two years apart. In total we analysed 857 brain scans. The data showed that grey matter volume in all four regions of the mentalizing network increases from age 7 until around age 9 or 10, when it reaches its highest point. The precise age at which this peak in grey matter occurs depends on the brain region, but it happens at some point in late childhood in all four mentalizing regions. After the peak, grey matter in all four regions gradually declines across adolescence and into the twenties. This protracted development demonstrates that areas of the brain involved in deciphering the mental states of others are still maturing anatomically from late childhood, through adolescence and into early adulthood.

∞

So, the mentalizing network develops in terms of its *structure* during adolescence. How does this collection of brain regions *function* when adolescents mentalize? A number of fMRI studies have examined this question and have revealed interesting differences in the way the network functions in adults and adolescents. A brain-imaging study carried out by Mirella Dapretto's research team at the University of California in Los Angeles involved scanning people's brains while they thought about what people mean when they say certain phrases. Children and young adolescents aged 9–14 years and adults aged 23–33 years looked at a series of cartoons and judged whether a phrase associated with each cartoon (for example, 'Nice going!' in the one reproduced overleaf) was sincere or ironic. Sometimes, the phrase was serious, for example when one of the children in the cartoon had inflated a balloon, and sometimes it was sarcastic, for example when she'd inflated the balloon so much that it had popped.

Young people aged 9–14 years and adults were both able to perform the task accurately – they were able to work out whether the phrases were sincere or sarcastic, which involves figuring out the intention of the speaker. However, the 9–14-year-olds showed more activity than the adults in one of the mentalizing regions – the dmPFC. In contrast, the adults activated other regions, including the superior temporal gyrus and fusiform gyrus, more than the children and

Sincere or ironic?

adolescents. This was one of the first fMRI studies to show that the pattern of brain activity associated with mentalizing – in this case, the ability to understand subtle meanings of speech in a social encounter – changes between late childhood and adulthood.

The results of this study are reminiscent of the findings from the developmental brain-imaging studies of self-processing described in chapter 2. Both this mentalizing study and those self-processing studies revealed that, as we get older, activity moves from the social brain region in the front of the brain (the dmPFC) to social brain regions towards the back of the brain (in the temporal cortex). In other words, it's not that adults and adolescents use entirely different networks of brain regions to do these tasks; rather, the same network is used, but the pattern of activity within the network changes.

The finding that dmPFC activity during mentalizing tasks decreases between adolescence and adulthood has been reported in studies investigating other aspects of mentalizing. A similar result was found in an fMRI study carried out by Stephanie Burnett-Heyes in my lab, in which we investigated social emotions, such as guilt and embarrassment. We refer to these as social emotions because they involve an understanding of someone else's mind – they require mentalizing. In order to feel guilty or embarrassed, you need to think about how someone else would think or feel about you in a given situation. These social emotions are different from basic emotions such as joy, fear, disgust or anger, which don't require you to think about another person's thoughts or feelings.

Imagine you have been daydreaming in a meeting and are asked a question, and your answer makes it very clear you haven't been listening. This is quite embarrassing, as you imagine what everyone else in the room must be thinking about you. Or you might have been talking negatively about someone who overheard what you were saying – in which case you'd feel guilty because you know they'd be hurt or offended by what you said. In both these situations, you are reading the minds of other people and inferring their mental states.

Our study investigated developmental changes in the way the brain processes social emotions. Adult women aged 22–32 and adolescent girls aged 10–18 read scenarios that evoked either social emotions (guilt or embarrassment) or basic emotions (fear or disgust). We had fun coming up with these! Here are a few examples of embarrassment stories, which were designed to be relevant to girls and young women: 'Your dad started doing rock 'n' roll dancing in the supermarket'; 'You fell asleep on the train and dribbled on the person next to you'; and 'Your friend told you that you had really bad breath.' And here are some guilt scenarios: 'You laughed when your friend told you she was feeling upset'; 'You kept some money that a little old lady had dropped in the street'; and 'You sent a really horrible text message to your friend when you were in a mood.' An example of a disgust sentence was 'You saw a pile of rotting guts near the dustbin at your friend's house,' and a fear sentence: 'You were sleeping in your friend's bedroom and a creature ran across your face.' We wanted to see what happened in the brain when the participants read these sentences aimed at arousing different emotions.

As expected, the social brain network was activated more when participants read the 'social emotion' sentences provoking guilt and embarrassment than when they read the 'basic emotion' sentences. This was the case in both age groups – but was there any difference between adolescents and adults? Echoing the findings of Dapretto's mentalizing study, our results showed that activity in the dmPFC when reading about social emotions was higher in the adolescent group than in the adult group. At the same time, another social brain region – the left anterior temporal cortex – was more active in adults than in adolescents when thinking about social emotions.

What accounts for this shift in activation pattern from adolescence to adulthood? Importantly, there was no difference in the degree of guilt or embarrassment elicited by the scenarios between the age groups. There are at least two possible explanations. One is that this *functional* change with age is associated with *structural* changes that occur during this period. As the study by Kate Mills described above showed, the grey matter in the brain is declining (structural change) at the same time as the decline in dmPFC activity (functional change) is occurring. Some scientists have suggested that the two processes might be related.

The second possible explanation is that the cognitive *strategy* for mentalizing changes between adolescence and adulthood. Perhaps when thinking about other people's minds, adolescents rely more heavily than adults on a simulation-based strategy in which they imagine how they would feel in a particular situation.

The dmPFC is involved in many cognitive processes, including the ability to think consciously about the self and other people. In contrast, the temporal regions of the social brain, used more by adults when mentalizing, are involved in storing social memories, such as your last birthday party or a cinema trip with friends. These social memories are drawn upon to construct 'social scripts' for all sorts of situations. A social script is a series of actions and outcomes that are common to a particular situation, so that when you next encounter the situation, you know what to expect. Take, for example, a trip to the cinema: you meet your friends, collect your tickets, queue for popcorn and so on. This social script has been developed and fine-tuned with experience, and it enables you to predict what you will encounter next time you go to the cinema. The brain stores social scripts for many situations, so that thinking about social situations becomes more automatic and less demanding as you repeatedly encounter them.

If adolescents and adults do use different *neurocognitive strategies* when performing the same task, this would result in different patterns of brain activity. When reflecting on social situations, perhaps adolescents – who have less life experience than adults, and therefore fewer previous situations to draw on – consciously bring to mind and think through specific examples of previous experiences, entailing more activity in the dmPFC. For adults, who are able to scan their stored collection of social scripts, accumulated over years of experiences, the process of thinking about themselves in different social situations might be more automatic, involving little conscious awareness. That might result in higher levels of activity in the temporal regions, which store social memories and social scripts, and lower levels of activity in the dmPFC, in adults.

One prediction arising from this proposal is that social tasks might interfere more with other tasks carried out at the same time in adolescents than in adults. Evidence to support this suggestion comes from a study in which we investigated adolescents' and adults' ability simultaneously to carry out a social cognition task and a working memory task. Adolescents (aged 11–17 years) and adults (22–30 years) were asked to keep track of non-social information (remembering three two-digit numbers) while carrying out a perspective-taking task (the Director Task). Performing these tasks at the same time was harder and incurred a performance cost in both age groups but, overall, adults were more adept at multi-tasking than adolescents. This supports the notion that, with age, social processing might become more automatic and so interfere less with other cognitive processes.

Research has demonstrated, then, that the social brain develops both structurally and functionally during adolescence: structurally in the decrease of grey matter volume between childhood and the early twenties, and functionally as activity within the mentalizing network shifts from regions at the front of the social brain to regions at the back of the social brain.

∞

So far in this chapter, I have focused on mentalizing. But understanding other people involves many social processes in addition to mentalizing. What about other aspects of social cognition; how do they develop in the brain? Adolescents are exposed to many more faces than are children, and it is during this period that we start to assess faces for different properties, such as sexual attractiveness, whether the person is a friend or not, and their social status. In the previous chapter I described how the ability to perceive and remember faces changes in adolescence. Is this because of changes in the brain? We know that the ability to recognize a face is present at birth, and that this very early face recognition probably depends on subcortical regions. After the first few months of life, regions in the cortex on the surface of the brain start to take over face processing.

One region that appears to be particularly specialized for faces is located towards the back of the brain in the fusiform gyrus – so much so that it has become known as the fusiform face area (FFA). The FFA is activated in brain-scanning studies whenever participants look at photos or cartoons of faces, and damage to this region causes people to have problems recognizing faces. This difficulty in recognizing faces, when other aspects of visual processing are normal, is called *prosopagnosia* or face-blindness.

Prosopagnosia doesn't only occur after damage to the FFA. People with *developmental prosopagnosia* are unable to recognize faces, whereas they are able to recognize objects and other visual stimuli. It has been estimated that up to 2.5 per cent of the population has prosopagnosia to some degree.*

A famous case of prosopagnosia was the neurologist Oliver Sacks, who, although he knew all about the condition and wrote about it in his book *The man who mistook his wife for a hat*, didn't realize until later in life that he himself was face-blind. He could see perfectly well and recognize objects and people, but he couldn't recognize familiar faces – not even that of his assistant, with whom he had been working for six years. He writes:

> I had been working with my assistant, Kate, for about six years when we arranged to rendezvous in a midtown office for a meeting with my publisher. I arrived and announced myself to the receptionist, but failed to note that Kate had already arrived and was sitting in the waiting area. That is, I saw a young woman there but did not realize that it was her. After about five minutes, smiling, she said, 'Hello, Oliver. I was wondering how long it would take you to recognize me.'

Brad Duchaine and other scientists working on developmental prosopagnosia have found that the FFA – along with other regions of the face-processing network – responds differently in people who are face-blind. However, other studies have called into question the notion that the FFA is specialized for faces alone. In one fMRI study, car and bird experts were scanned, and there was activation in the FFA when car experts identified cars and when bird experts identified birds. This suggests that the FFA is involved in expertise, and specifically in the visual recognition of anything in which someone is an expert – for

* Brad Duchaine, at Dartmouth College in the United States, has studied face recognition for many years and is particularly interested in prosopagnosia. He has developed a face-recognition test, which you can carry out online to assess your own face-recognition skills: http://www.faceblind.org/facetests/.

most of us this will be faces, but for some people it will extend to other categories of object.

Perceiving and remembering faces involves other regions of the cortex too, including areas in the occipital gyrus and the superior temporal sulcus (STS). Many regions within the core face-processing network continue to develop between adolescence and adulthood. The FFA becomes gradually specialized for faces across childhood and adolescence, and at the same time we become better at remembering the identity of faces. In a study by Suzanne Scherf and colleagues at the University of Pittsburgh, children, adolescents and adults viewed short movie clips of faces, places and objects while having their brains scanned. The study showed that activation in both the FFA and the STS became increasingly selective for faces with age, meaning that these regions start to respond more to faces and less to other objects.

Studies by Kathrin Cohen Kadosh in my lab showed that the face-processing network becomes specialized during adolescence, particularly in relation to complex aspects of face processing such as extracting socially relevant information, including recognizing someone's identity or their emotional expression. In our fMRI study, we scanned 48 children, adolescents and young adults aged between 7 and 37 years while they carried out three different face-processing tasks: detecting the identity of a face, its emotional expression (angry or happy, for example), or the direction of its eye-gaze. All age groups activated the core face-processing network during each task. However, as age increased, additional brain regions became activated for each of the three tasks – and these were slightly different regions for each task. This supports the idea that the face-processing network becomes increasingly specialized with age.

In a different study by Christopher Monk, Danny Pine and colleagues at the NIMH in the United States, young people aged 9–17 and adults aged 25–36 viewed fearful and neutral faces. Fear in a face is identified mainly in the eye region – to recognize that someone is looking scared, you need to notice that their eyes are open wider than usual and the whites of their eyes are showing. Participants in this fMRI study were instructed in one iteration simply to look at fearful faces and neutral faces, or in another iteration to focus their attention on either the emotional facial feature (the eyes) or a non-emotional facial feature (the nose).

Compared with adults, the children and adolescents showed higher activation of two regions in the frontal cortex when they were asked to look at fearful faces than when they were asked to look at neutral faces. When they were instructed to focus their attention on a non-emotional aspect (the nose) of fearful faces, activity in one of the frontal regions (an area called the anterior cingulate cortex) was higher in adolescents than in adults. This indicates that, in adults, brain activity changes according to what participants are asked to focus their attention on. This makes sense – it's important to be able to focus on some things in your environment but not others; giving equal attention to everything would overload the system and be overwhelming to process and interpret. But in adolescents, Monk and his colleagues found, brain activity is modulated according to the emotional nature of a stimulus, irrespective of what the participants are told to focus on. This suggests that the adolescent brain is tracking emotional and arousing stimuli in the environment even when the individual has been asked to focus on non-emotional stimuli.

Thus it seems that the way the brain enables us to pay attention to a non-salient object (in this case, the nose of a fearful face) in the presence of something attention-grabbing but irrelevant (the eyes of a fearful face) is still undergoing development between adolescence and adulthood. This fits with the possibility that adolescents find it challenging to focus on a task at hand in the presence of emotionally salient and distracting stimuli, such as driving a car with a friend in the passenger seat. And this takes us back to one of the most persistent and challenging features of adolescent behaviour: the penchant for risk-taking.

9

The right sort of risks

SMOKING, BINGE-DRINKING, EXPERIMENTING WITH drugs, unsafe sex, dangerous driving – all are more common in adolescents than in adults. Society stereotypes adolescents, and particularly adolescent boys, as reckless risk-takers. But is it as simple as this? Do all adolescents take risks? In some contexts adolescents actually avoid risk-taking. Teachers often have to encourage young people to ask or answer questions in class, guess an answer in a test, try new lines of argument and so on. These are risks, and apparently adolescents don't like taking them. And, when adolescents do take risks, there's probably good reason for it.

That's not to say that risk-taking isn't a serious problem: the leading cause of death in adolescence and young adulthood in Western countries is accidents, and these are sometimes a result of risk-taking – primarily reckless driving. We should bear in mind, of course, that other causes of death for this age group are rare: adolescents are less likely than either children or adults to die from health-related causes. At the same time, they *are* more likely to take risks that result in accidents, even fatal accidents.* Ron Dahl, from the University of California in Berkeley, has referred to this as the 'paradox of

* It's interesting and important to note that the proportion of deaths as a result of accidents among young people aged 15–19 in the United States has actually fallen quite dramatically over the past two decades. Is this because modern adolescents are, as a group, less likely to take risks?

adolescence': in the period of life during which people are at their healthiest and fittest, there is still mortality, caused mostly by accidents that are, in principle, largely preventable.

However, the picture is more complicated than the stereotype of the reckless and thrill-seeking adolescent suggests. First, while risk-taking in adolescence can lead to injury and illness, including long-term problems associated with smoking, drinking, taking drugs and committing crime, it is worth noting that death is fortunately rare, with survival rates of North American high-school students at over 99.5 per cent. The risks most adolescents take do not result in serious harm – to themselves or other people. Even for less extreme risks, there are large individual differences: some individuals are risk-takers, while others are not.

It is also important to consider wider contextual factors that enable risk-taking in adolescence – notably the increased freedom permitted by parents and society. Adolescents are given more independence than children, spend more time unsupervised, and are allowed and indeed encouraged to make their own decisions – all opening up opportunities for increased exploration and risk-taking. In contrast, parents typically set boundaries and constraints on the decisions of younger children, who are not able to take as many risks as they might otherwise.

That said, policy-makers and parents alike do worry about the risks adolescents take, and there's quite a lot of research on the causes and consequences of risk-taking in adolescence.

Laurence Steinberg has carried out many studies on risk-taking in this age group. In his excellent book *Age of opportunity*, Steinberg describes his compelling research on adolescence and his work with young people being prosecuted in the American legal system. Steinberg has acted as a defence witness in cases in which adolescents have been prosecuted for dangerous decisions they have made, such as driving risks that have gone wrong. He argues that brain development, and the fact that risk-taking is a natural part of adolescent development, must be taken into account when considering whether an adolescent is guilty of a crime or not.

In the late 2000s Steinberg, along with other researchers including B. J. Casey, proposed a theory to explain why risk-taking peaks in adolescence. Both Steinberg's and Casey's theories involve the brain's limbic system, which (among other processes) generates the rewarding feeling – the kick – elicited by taking risks. The core idea is that, in young adolescents, the limbic system is already mature and particularly sensitive to the rewarding feeling that risk-taking sometimes elicits. At the same time, the prefrontal cortex – which stops us acting on impulse and inhibits risk-taking – is not yet mature, and will continue developing throughout adolescence and early adulthood.

The theory suggests that this results in a 'developmental mismatch' between the maturity and functioning of these two brain systems, and this in turn explains why adolescents get a kick out of taking risks (a function of the limbic system) and aren't always able to stop themselves doing so in the heat of the moment (a skill that relies on the prefrontal cortex). In contrast, the theory suggests, adults are better at regulating behaviour and stopping themselves taking dangerous risks, even when the risks are exciting and potentially rewarding, because of their mature prefrontal cortex. Steinberg called this the 'dual systems model' because of the two brain systems involved.*

These theories are based on the assumption that the brain's reward and emotion systems mature earlier in adolescence than the prefrontal cortex control system. What is the evidence for this? Until recently there wasn't much. So in 2014 Kate Mills, Anne-Lise Goddings and I carried

* Several versions of this theory have been proposed by other researchers, such as the 'imbalance model' put forward by B. J. Casey and Leah Somerville at the Sackler Institute in New York. Very similar to the dual systems theory, the imbalance model suggests that the developmental mismatch between the brain's reward system and the control system explains the adolescent tendency to take risks and act impulsively.

out an analysis in collaboration with Jay Giedd from the National Institute of Health in Bethesda to look at the question in a bit more detail.

We returned to the MRI data from Jay Giedd and Judith Rapoport's large developmental MRI study (see chapter 6). However, we weren't able to include all the participants' data in our analysis. This is because we needed MRI data from people who had been scanned at least three times, including once in late childhood, once in mid-adolescence and once in late adolescence/early adulthood, and not many people in Giedd's sample had had at least three scans covering all these age bands. Furthermore, because we were interested in the development of the limbic system, we needed brain scans that included high-quality images of these subcortical structures. That's a challenge, as these small structures deep inside the brain (see illustration in chapter 4) can be blurred and distorted in MRI scans.

Kate and Anne-Lise, who were studying for their PhDs at the time, went through each and every scan by eye to check the image quality of the limbic system. This was meticulous work that took many weeks. PhD research can be gruelling! Many scans were not of sufficiently high quality to enable us to analyse the limbic structures, but we were able to obtain multiple high-quality MRI scans from 33 individuals.

We were interested in whether the limbic regions – specifically the nucleus accumbens, which processes reward, and the amygdala, which processes emotion – reach maturity earlier than the prefrontal cortex. We analysed each of these regions in terms of the amount of grey matter it contained. In our study, a brain region was defined as 'mature' when its grey matter volume appeared to have stopped changing.*

We first carried out an analysis in which we put all the partici-

* Assessing when grey matter stops decreasing and levels out is only one possible way of defining brain maturity. In a 2016 paper, Leah Somerville discusses the many ways in which brain maturation could, in theory, be defined, including the plateauing of developmental changes in structure, function, connections, neurochemicals and behaviour.

pants' data together and analysed the average development of grey matter volume in the two limbic regions and the prefrontal cortex across time. This analysis of *average* development, which is shown in the three graphs on the left-hand side of the page overleaf, showed that grey matter in the amygdala increases by about 7 per cent throughout late childhood and early adolescence, and stops changing much after age 14. Grey matter in the nucleus accumbens slowly declines throughout late childhood and adolescence, losing about 7 per cent of volume during this period. In contrast, grey matter in the prefrontal cortex declines dramatically between late childhood and early adulthood, reducing by around 17 per cent.

A look at these graphs suggests that the limbic regions undergo a different pattern of development from the prefrontal cortex. The amygdala does indeed appear to mature earlier than the prefrontal cortex, which undergoes substantial and protracted change throughout adolescence and into early adulthood. The nucleus accumbens also changes over this age range – but far less than the prefrontal cortex. This analysis of average development in the three brain regions thus supports the general notion that there is a mismatch between the development of the limbic system – in this case the amygdala, and to a lesser extent the nucleus accumbens – and the prefrontal cortex.

We then carried out a different analysis in which, instead of calculating averages from all the participants' MRI data, we looked at each individual's brain development separately. This painted quite a different picture, as you can see from the right-hand trio of graphs overleaf. Instead of each participant showing the same pattern, there were large individual differences in the development of each brain region.

Kate, Anne-Lise and I independently looked at graphs of the developmental pattern of the three regions in each individual to determine whether one region matured before the others. We did this separately, so none of us would be biased by what the others thought, and also blindly – we didn't know which region was which on the graphs.

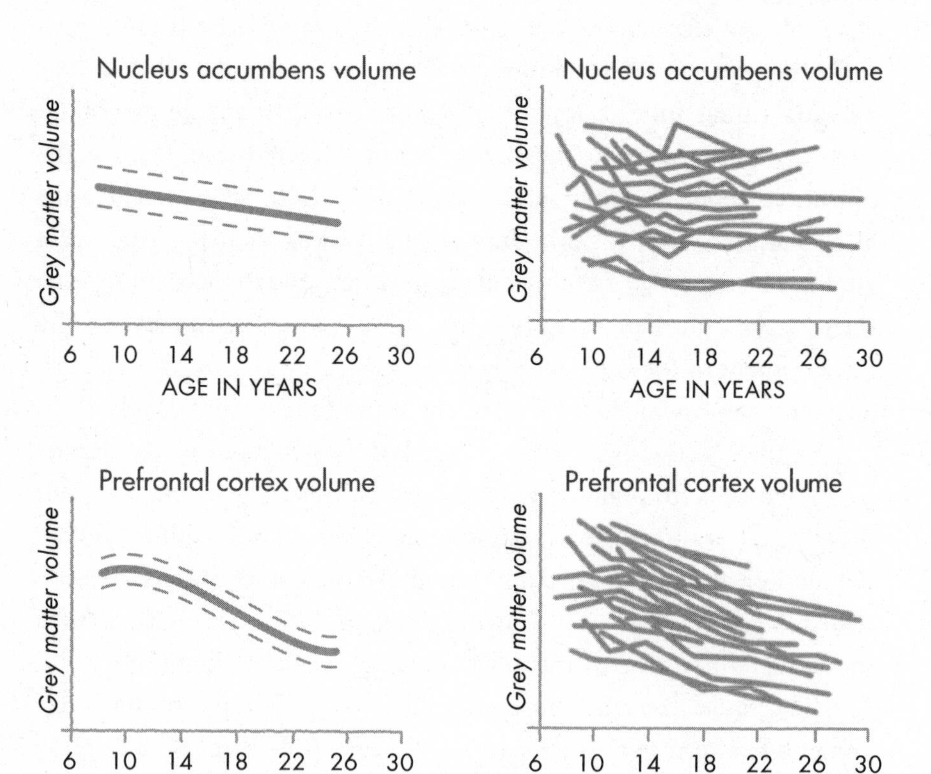

Grey matter development in three brain regions: average and individual raw data (dotted lines in average data represent confidence intervals)

This visual inspection of the graphs by all three of us confirmed that there were large differences between individuals in terms of the development of the limbic system and prefrontal cortex. Of the thirty-three participants, fifteen were considered to show a complete mismatch, in which both limbic regions (the amygdala and the nucleus accumbens) reached maturity before the prefrontal cortex. In twelve participants, the amygdala (but not the nucleus accumbens) was considered to have matured before the prefrontal cortex. In two participants, the nucleus accumbens (but not the amygdala) was considered to have matured before the prefrontal cortex. And in the remaining four participants, there was considered to be no mismatch at all – in these people's brains, all three regions had matured at around the same age.

What does this mean? One important conclusion is that individual differences are just as significant as – perhaps even more important than – averages. This is an important point, because most studies of brain development (indeed, most scientific studies of people) focus on averages. Many researchers in the field of adolescent brain development are now recognizing that it's time to move beyond averages and start looking at individual differences. After all, there's no average teenager.

What causes these individual differences, and what are their consequences? What's the difference between a boy whose amygdala and nucleus accumbens develop much earlier than his prefrontal cortex, and a boy whose limbic system and prefrontal cortex develop more or less at the same time? Is the first more of a risk-taker than the second? We would love to know the answer to this question, but unfortunately it's not so easy to determine from the MRI data we had. Jay Giedd's MRI study started in 1991, and he wasn't specifically interested in risk-taking. He suspected he'd find that, in general, the brain changes throughout development more than was previously assumed. But he couldn't have known that, fifteen years after his study began, major theories of adolescent risk-taking would be based on the precise developmental timings of different brain regions. In fact, these theories were partly based on Giedd's own data showing slow, protracted development of the prefrontal cortex across adolescence and into early adulthood. If he had known how much the brain develops in adolescence before his study started, perhaps he would have asked participants about how many risks they took and how impulsive they were – that way, we could have explored how each person's risk-taking habits were linked to their brain development. As it is, we don't have this information from Giedd's study.

In our study, we did send questionnaires about risk-taking and impulsivity to the participants, who were in their thirties by that point. We asked them to try to remember how many risks they took as teenagers.* We asked them to write down some of the risks they took in their 'riskiest year' as an adolescent to jog their memories. These questions aren't that easy to answer even if you're thinking about your current behaviour, let alone when you're trying to remember what you were like many years ago. In other words, this is not an ideal way to assess adolescent risk-taking, but it's the best we could do.†

The questionnaire data didn't explain anything – there was no link between whether participants considered themselves risk-takers or not and their brain development. But this might have a good deal to do with the less-than-ideal methods of assessing risk-taking. Fortunately,

* For example, in one part of our questionnaire (designed to measure sensation-seeking), participants were asked to rate the following as True or False: (1) I liked to have new and exciting experiences and sensations even if they were a little frightening; (2) I liked doing things just for the thrill; (3) I liked to do things that were a little frightening; (4) I would try anything once; (5) I did 'crazy' things just for fun; and (6) I liked wild and uninhibited parties. Each 'true' statement was counted, so that participants could have scores between 0 and 6.

† Some of the risks participants wrote down were amusing, including 'dangerous gymnastics'. I also discovered that American teenagers from the 1990s (in Jay Giedd's cohort) took different risks from the risks my British friends and I took as teenagers. They seemed to be particularly into something called 'toilet papering', in which you wrap neighbours' trees and houses in toilet paper.

in new studies of brain development, scientists are including excellent measures of risk-taking and impulsivity at the same time as acquiring brain scans. We need this knowledge to understand more about differences in brain development between individuals, and to find out whether the dual systems theory of risk-taking is correct, at least for those adolescents whose limbic system and prefrontal cortex are maturing at different rates.

Another question is what happens to these different people when they become adults. What is the consequence of having a limbic system that develops earlier than the prefrontal cortex? Does this result in a different adult from someone whose limbic system and prefrontal cortex develop in tandem? Again, we don't know the answer to this fascinating question – yet. And again, this is the kind of question that will surely be answered in future studies of brain development. The different rates and patterns of structural change in different people's brains certainly suggests that there is a consequence to be discovered.

∞

The dual systems model also suggests that the limbic regions are *hypersensitive* in adolescence, meaning they are more active during risk-taking and reward-processing in adolescents compared with adults. There is quite a lot of evidence that, in adolescence, the limbic system is particularly sensitive to rewards (money is often used as the reward in these experiments) that are obtained as a result of risk-taking. This evidence comes from fMRI studies that have measured brain activity during risk-taking and reward-processing. Many of these have employed gambling tasks in which the participant makes a decision that results in winning or losing money.

One of the earliest studies of this kind was carried out by Monique Ernst and her colleagues at the NIMH in Bethesda. Adolescents and adults played a 'Wheel of Fortune' task, in which you spin a wheel and the number it lands on determines whether you win a small or large amount of money, or nothing at all. When adolescents won small or large amounts of money on the task, there was more activity in the nucleus accumbens (part of the limbic system that processes reward) than in adults.

Subsequent fMRI studies have supported the finding that, compared with adults and children, adolescents show different neural responses to reward. One study, conducted by Eveline Crone and her colleagues in Leiden, the Netherlands, measured brain responses when participants aged between 8 and 26 years played a child-friendly gambling task. On each trial, the participant is presented with a choice between two types of gamble. Some gambles were low-risk, with a high probability of obtaining a small reward (1 euro). Others were high-risk gambles, with a much smaller probability of obtaining a higher reward (2, 4, 6 or 8 euros) than in the low-risk gambles.

When participants chose high-risk gambles, regions in the prefrontal cortex associated with self-control showed a steady decrease in activity with age,* while reward-related regions (the nucleus accumbens and ventromedial prefrontal cortex) showed a peak in activity in mid- to late adolescence. Again, this result fits with the notion that hypersensitivity of reward-processing regions, together with a steadier and slower development of the prefrontal cortex, might underlie adolescent risk-taking.

It's important to note that just because regions such as the nucleus accumbens are more active in adolescence, that doesn't mean that adolescents experience rewarding feelings more strongly. We cannot draw any conclusions about what is going on at a psychological or cognitive level on the basis of measures of brain activity – this is a fallacy called 'reverse inference'. All brain regions are involved in multiple cognitive, perceptual, emotional and/or motor processes, and activation in one region might be attributable to any number of

* Note that development can involve increases or decreases in activation in certain brain regions.

psychological experiences. The only way to know if something *feels* more rewarding in adolescence is to ask participants, which is rarely done in studies of risk-taking. Activity in the nucleus accumbens is not in itself evidence of rewarding feelings.

Although many studies have shown that areas of the brain associated with reward are more highly activated in adolescents than in other age groups, as with most neuroimaging studies there are some inconsistent results, with a small number of studies not replicating this developmental difference or even showing the opposite pattern. These discrepancies might arise because studies have used slightly different tasks and included slightly different age groups.

Heightened risk-taking and novelty-seeking in adolescence are not specific to humans: they are also present in other animals, including rats. There is also evidence that reward is processed differently in the brain's reward system in adolescent compared to adult rats. One study showed that a larger proportion of neurons in the dorsal striatum (a reward region) were activated in anticipation of reward in adolescent rats than in adult rats.

The evidence of hypersensitivity of reward regions to rewarding stimuli, taken together with the relatively slow development of prefrontal regions involved in self-control, certainly lends support to the dual systems theory of adolescent development. B. J. Casey and Leah Somerville have suggested that this might especially be the case when decisions are made in an emotional – 'hot' – context. Imagine you're deciding whether or not to have a third alcoholic drink. If you're on your own at home you might think of all the reasons this isn't a good idea and stop yourself pouring another. If you're making the same decision in a bar with your friends, who are all having another drink, your decision might go the other way. This is the difference between a cold and a hot context.

A task that has been used to look at this aspect of decision-making is called an 'emotional go/no-go' task. This is similar to the go/no-go task I described earlier, but in the emotional version participants are presented with a series of different faces. Most faces show a happy expression, but occasionally a face with a neutral expression appears on the screen.

Participants are asked to respond to the faces by pressing a key, but only when the face on the screen is happy (the 'go' stimulus). If the face is neutral, participants must refrain from pressing anything (the 'no-go' stimulus). This is surprisingly challenging – you get into a rhythm of tapping a key every time a face appears, and have to make a concerted effort to stop this automatic response when the face that appears is neutral. In the second version of the task, the faces are switched: the neutral faces are the go stimuli and are more numerous than the happy faces, which are the no-go stimuli. In go/no-go tasks, researchers record how many times participants accidentally press the button in response to a no-go stimulus as a measure of impulsivity.

In a study by Casey, Somerville and Todd Hare, a group of participants aged 6 to 29 carried out this task while lying in an fMRI scanner. There were a number of interesting findings. First, the ability to inhibit pressing the button when the no-go stimulus was a neutral face improved steadily with age. In parallel, activity in the prefrontal cortex (possibly underlying inhibitory control) gradually increased with age. In contrast, compared with children and adults, adolescents showed a reduced ability to inhibit pressing the key *when the no-go stimuli were happy faces*. In other words, the ability to stop yourself making an automatic response to *positive* stimuli showed a dip in adolescence. In parallel, there was a peak in activity in the ventral striatum (a reward area) in adolescence in this version of the task. The authors suggested that adolescents show enhanced sensitivity to rewarding stimuli – in this case positive faces, which are more pleasing than neutral faces. This, they suggest, might contribute to heightened risk-taking, as adolescents are more compelled to act impulsively to gain rewards, even when they're trying not to.

∞

When we make risky decisions we sometimes regret them, while at other times we might be relieved things worked out the way they did. Emotions such as regret and relief are *counterfactual emotions*, as they involve thinking about what could have happened as a consequence of a decision, but did not. They play an important role in decision-making and learning from previous mistakes. When we make a decision, we anticipate, and try to avoid, the undesirable feeling of regret based on past experience when we were faced with similar choices. Abigail Baird of Vassar College in the United States has suggested that the ability to think counterfactually about the outcomes of decisions continues to develop during adolescence. This might increase the likelihood of adolescents taking risks, as it would mean they are less likely to weigh up the possible outcomes based on past experience of counterfactual emotions.

Stephanie Burnett-Heyes and I, together with collaborators including Giorgio Coricelli, based at the University of Southern California, investigated this possibility by looking at the development of counterfactual emotions in making decisions. We used a gambling task that Giorgio had developed, and tested participants aged 9–35 years. On each trial, participants chose between two gambles, presented as two pie charts on a computer screen. The pie charts indicated how likely the participant was to win or lose money. Faced with the screen shown in the illustration on the next page, the participant could choose either a gamble with a small chance of winning 200 points and a large chance of losing 200 points (A), or a gamble with a 50:50 chance of winning or losing 40 points (B). When the participant had made their choice, a pointer spun around the pie chart they had chosen, and they were then shown how many points they had won or lost. After seeing the results of each gamble, participants were then asked to rate how relieved or regretful they felt about the choice they had made.

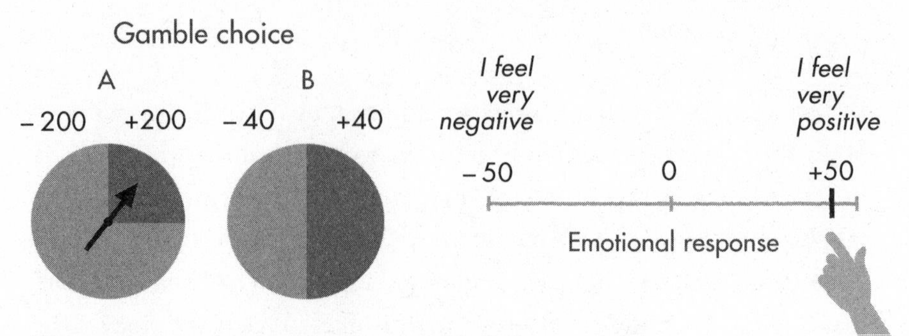

The gambling task: a teenager is more likely than an adult to choose the risky option (A), and here the gamble pays off.

The results showed that risk-taking – choosing the gamble with a small chance of winning a lot over the one with a good chance of winning a moderate amount – peaked in mid-adolescence. We calculated the age at which our group of participants made the greatest proportion of risky choices to be 14.38 years.*

We also found that the strength of counterfactual emotions increased between childhood and adolescence: on learning the outcome of each gambling decision, mid-adolescents aged 12–15 years felt a stronger sense of relief (and, to a lesser extent, regret) than did 9–11-year-olds. The late adolescent and adult groups' emotional responses were no higher than those of the mid-adolescent group. This suggests that there may be a relationship between the intensity of feelings of relief (and regret) and the propensity to take risks, both of which are higher in mid-adolescence than earlier in childhood.

Our task, in which participants had to evaluate how they felt as a result of each gamble, can be considered a 'hot' decision-making task

* Having said that, it makes little sense to try to pinpoint the specific age at which a certain behaviour peaks, as there are such strong individual differences. Some people might take most risks at age 14, others at age 24; others still might show no obvious peak in risk-taking.

– feelings are involved in the decisions, and emotions are raised. The evidence from our study, and many others, suggests that adolescents are more likely than children and adults to make risky decisions in 'hot' contexts.

Psychologists often use card games to measure impulsivity and risk-taking in hot contexts, when money is at stake. A 2010 study by Elizabeth Cauffman, Laurence Steinberg and colleagues used a version of the Iowa Gambling Task. This task requires participants to choose among several packs of cards, each associated with a different likelihood of winning or losing money. Some packs seem to be lucrative, giving large payouts, but eventually result in catastrophic loss because they also contain cards associated with large losses – you can think of these as the risky decks. Other packs are 'steady earners', with small wins hardly ever being offset by even smaller losses. Participants can play cards from each deck over several rounds, and will usually start to show a preference for one of the packs. The Iowa Gambling Task has been used in many different studies, and these have shown that healthy adults tend to sample the risky packs initially, but then settle on the safer options.

In Cauffman's study, the task was completed by 901 participants aged between 10 and 30 years. The tendency to play increasingly from the risky deck was strongest in adolescence/early adulthood (14–21 years): this age group preferred and persisted with the risky pack, even though it eventually lost them money. In contrast, children and adults were more likely to learn to avoid this pack over the course of the game. It seems, then, that adolescents' and young adults' behaviour is biased towards choices that might result in high rewards, even when this behaviour may have negative consequences. This suggests that the ability to use information about monetary value and the probability of winning or losing to guide decisions in 'hot' contexts, characterized by money, high emotion or arousal, is still developing in adolescence. Another 'hot' context is when adolescents are with their friends, which, as we saw in previous chapters, is associated with increased risk-taking in adolescence.

In contrast, in 'cold' tasks, with no emotional context, risk-taking isn't necessarily increased in adolescence. This is an important distinction, because it challenges the notion that adolescents always take risks. This is simply not true – some adolescents take more risks in certain situations, but not in others. If an adolescent is on her own, maybe walking back home from school and focusing on what she needs to do when she gets back, and is offered Ecstasy, she's much less likely to accept it than if she is at a party on a Saturday night and all her friends are taking the drug. This is an example of a different decision being made by the same person in a cold and a hot context. Of course, this does not change completely in adulthood – we are all influenced by our friends, and adults are probably also more likely to drink and take drugs when with friends than when alone.

There is some evidence that adolescents show a greater preference than adults for immediate rewards, which might lead them to take more risks. In a 'delay discounting task', a participant is given the choice between a small, immediate reward and a larger, delayed one. For example, would you rather be given £5 now or £30 in one month? What about £25 now or £30 in one month? What about £5 now or £30 in one year? As you can see, the amounts offered can be changed, and so can the length of the delay imposed on the second, larger amount. There are large individual differences in the tendency to act 'impulsively' in this task, with some people always opting for the smaller, immediate reward, while others always opt for the larger reward even if it means waiting a long time.

As impulse control gradually improves between childhood and early adulthood, the tendency to choose the immediate reward decreases. Brain-scanning studies have shown that this reduction in choosing an immediate but smaller reward is associated with a steady increase between late childhood and early adulthood in activity in the ventromedial prefrontal cortex, and a decrease during the same period in activity in the ventral striatum.

∞

The delay discounting task is a version of a famous task given to young children by Walter Mischel in the late 1960s. Mischel's version is called the Marshmallow Test because it involved leaving a child aged around 4 years alone at a table with a marshmallow. The experimenter told the child that he would be leaving the room for fifteen minutes, and that if the child had not eaten the marshmallow by the time he came back, the child would be rewarded with two marshmallows.

Resisting temptation – or not . . . the Marshmallow Test

This is a test of 'delayed gratification', and there were intriguing individual differences in the amount of time children could wait before succumbing and eating the single marshmallow. In his original study, Mischel tested thirty-two children. Some of them ate the marshmallow immediately; some managed to resist until the experimenter came back and were rewarded with two marshmallows. Many resisted the marshmallow for a few minutes, but eventually gave in to the temptation. There are some fun videos of children trying to resist the temptation, showing all kinds of behaviours to avoid eating the marshmallow: some children sniff it and even lick it, but then sit on their hands or turn away so they can't touch or see the tempting sweet.* The key to resisting temptation in order to receive a larger reward after a delay appears to be the ability to divert attention away from the tantalizing present reward.†

Mischel's original study demonstrated that there are large differences between children in terms of the amount of time they are prepared to wait for a delayed reward. But he didn't stop there. He and many different collaborators monitored the children who had taken part in the original study, and other similar studies that he performed, as they grew up. Intriguingly, the ability to resist the marshmallow was correlated with all sorts of positive outcomes later in life. The longer they could 'delay gratification' as children, the higher they were rated as adolescents by their parents in terms of being interpersonally competent, able to concentrate and to exert self-control, and the better they did in their school exams. Children who waited longer at age 4 did better in a go/no-go task when they were adolescents – they were quicker at responding to the 'go' stimuli without making more erroneous responses to the 'no-go' stimuli. The go/no-go task measures self-control, and this finding demonstrates that pre-school children with better self-restraint become adolescents who also have more self-control.

A subset of the children who had taken part in the original experiment took part in an fMRI study when they were in their forties. This

* https://www.youtube.com/watch?v=Yo4WF3cSd9Q.

† A more recent study showed that whether children can resist the marshmallow depends partly on context. In environments in which children learned that the experimenter was reliable and kept his word – for example, bringing a new set of colouring pencils after two minutes, as promised – children were more likely to wait longer for a marshmallow than when he was unreliable. See https://www.ncbi.nlm.nih.gov/pubmed/23063236.

study was led by B. J. Casey, who scanned participants while they performed the emotional go/no-go task using happy and neutral faces, which I described above. For the participants who had been good at delaying gratification as children, the prefrontal cortex differentiated between no-go and go trials (regardless of facial expression) to a greater extent than in the participants who had not been so good at the self-control task at age 4. In contrast, in the participants who had not been so good at delaying gratification as children, the ventral striatum showed a higher response to the no-go stimuli when they were happy faces. Thus the reward system was more active in those individuals who were less able to delay gratification when they needed to resist alluring stimuli (happy faces). The researchers suggested that this might contribute to their lower ability to resist temptation. This series of studies indicates that the self-control as measured originally by the Marshmallow Test is a relatively stable individual attribute.*

Mischel's results introduced the idea of self-control as an important skill, a notion that has since been supported by many further studies. Terrie Moffitt and Avshalom Caspi, who work at Duke University in the United States and King's College London, have reported related findings from their longitudinal study based in Dunedin, New Zealand. They have been studying 1,037 children born in Dunedin between April 1972 and March 1973 for many years. These people are now in their forties and still being studied. Extraordinarily for this kind of longitudinal study, almost all of the original participants are still taking part in the research.

Many aspects of the children's development were assessed at several time points. One skill Moffitt and Caspi measured was self-control, which was assessed at the ages of 3, 5, 7, 9 and 11 years. They evaluated self-control by asking the children, their parents and their teachers to fill in questionnaires about how impulsive, hyperactive and good at

* Mischel has written a wonderful book about his research, entitled *The marshmallow test*.

self-regulation each child was. Moffitt and Caspi and their colleagues combined the results from all five time points into a single, composite measure of childhood self-control.

Intriguingly, the level of self-control in childhood predicted a range of outcomes in adolescence and adulthood. Children with lower self-control scores went on to have poorer health, including lower lung capacity and a higher prevalence of gum disease and obesity, at age 32. Children with lower self-control were also more likely to become dependent on drugs, such as tobacco, alcohol or cannabis, than children with higher self-control.

Self-control in childhood also predicted social and financial outcomes in adolescence and adulthood, as shown in the chart below. Children with lower self-control were more likely to leave school with

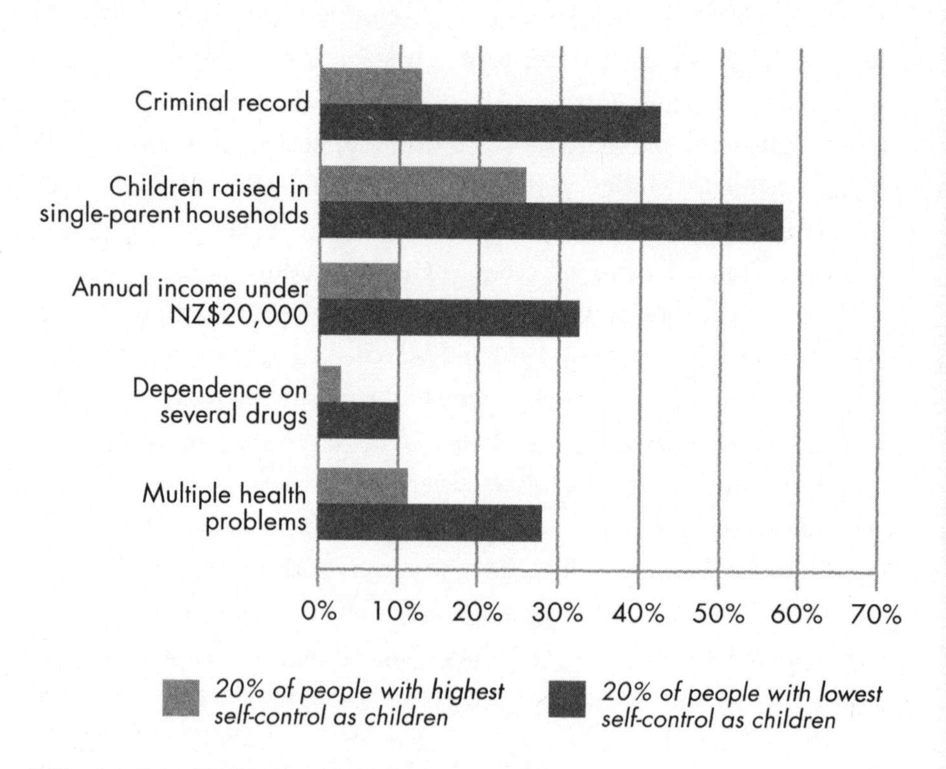

Self-control in childhood and outcomes in adulthood

no qualifications and have unplanned pregnancies as teenagers. As adults, they had higher levels of debt and fewer savings, and were more likely to have been convicted of a crime, than children with higher self-control.

Moffitt and Caspi's findings are *correlational* rather than *causal* – that is, they show *associations* between self-control and other factors, but not necessarily that certain outcomes occur *because of* high or low self-control. Nevertheless, they do suggest that self-control in childhood seems to be an important skill that permeates behaviour throughout life.

The good news is that self-control can, to a certain extent, be trained. There are lots of ways to do this, and different training methods suit different people. One possible way to improve self-control in some people is mindfulness training. Mindfulness is a state of being, cultivated by meditation, in which one focuses awareness on the present moment, while acknowledging and accepting one's thoughts, feelings and bodily sensations. Together with Mark Williams and Willem Kuyken from the University of Oxford, and Tim Dalgleish from the Medical Research Council Cognitive and Brain Sciences Unit in Cambridge, we are starting a very large trial to look at whether mindfulness training in schools has beneficial outcomes in terms of mental health and well-being in young people aged 11–16 years. Our trial involves more than eighty schools in the UK. The results will contribute evidence to the question of whether mindfulness training should be carried out in schools, as part of the social–emotional learning curriculum. If the results of our trial show that mindfulness training in schools significantly improves self-control and well-being in young people and reduces the risk of mental health problems such as depression and anxiety, then it might be a good idea to consider introducing it into all schools. But we won't know the answer for another six years – such a large trial takes a long time.

∞

How does adolescents' sensitivity to reward influence their behaviour in other contexts? A 2016 study in my lab revealed that adolescents are less likely to learn to avoid punishment. Our study, led by Stefano Palminteri, compared how adolescents and adults learn to make choices on the basis of the information available to them. A group of adolescents aged 12–17 years, and a group of young adults aged 18–32 years, completed tasks in which they had to choose between abstract symbols. Each symbol was consistently associated with a fixed chance of a reward (gaining money) or punishment (losing money). As the task progressed, participants were able to learn which symbols were rewarding and which were punishing, and gradually adjusted their choices accordingly.

The results showed that adolescents and adults were equally good at learning to choose symbols associated with reward, but that adolescents were worse than adults at avoiding symbols associated with punishment. Adults also performed significantly better when they were told after each choice what would have happened if they had chosen the other symbol, whereas adolescents did not appear to take this information into account. The results suggest that adolescents and adults learn in different ways, and that a reward-based approach, rather than punishment, might be more likely to be effective in adolescent learning.

∞

Where does the research described in this chapter take us? If adolescents tend to be more influenced by reward than by punishment, and are more tempted by immediate rewards than inclined to wait for larger rewards later, then perhaps this explains why some advertising warning about the long-term damaging effects of, say, smoking has little effect in this age group.

Anti-smoking advertising campaigns targeting young people have historically focused on the long-term health risks; but we know that this isn't going to have as much impact on young people as focusing on more immediate and non-health-related consequences. One study from 2000 compared the impact of the following different messages in many anti-smoking advertising campaigns aimed at teenagers:

- long-term health effects of smoking, such as cancer and lung disease;
- short-term cosmetic effects of smoking, such as smelly breath, and related outcomes such as romantic rejection;
- marketing practices, such as the use of models and the distribution of free promotional items;
- deceptive tactics used by tobacco companies to sell what is essentially a lethal and addictive product, such as illegally targeting minors and falsely claiming that cigarettes are non-addictive;
- second-hand smoke and its impact on family members and other people, particularly infants and children;
- smokers portrayed as negative role models;
- refusal skills – attractive, personable individuals shown refusing to smoke.

The researchers found that the advertisements that had the greatest potential for dissuading young people from smoking were ones that focused on second-hand smoke, the smoker as negative role model, refusal skills and the deceptive portrayal of a lethal product. None of these messages has anything whatsoever to do with the actual health risks to the smoker, but instead largely focused on the effects on other people.

A similar finding emerged from a study in which adolescents were told how companies are trying to manipulate them into unhealthy eating. In one intervention, healthy eating was framed as a way to take a stand against manipulative and unfair practices of the food industry, such as engineering junk food to make it addictive and marketing it to children. In another intervention, the focus was on traditional health education.

The group who had been told about the coercive and underhand practices of the food industry saw healthy eating as a way of exercising independence and taking a stand in line with social justice. This group were more likely than the other group to select healthier options and resist unhealthy snacks and drinks one day later.

It seems, then, that public health interventions aimed at adolescents are likely to be more effective when they focus on values that are important to young people: in this case, feeling like a socially conscious, autonomous person. This healthy eating intervention might also play into the adolescent penchant for rebelling against authority – in this example, the adult-run food industry.

Taken together, these studies suggest that, if we want to curb certain kinds of risk-taking in young people, it would be a good idea to focus on the immediate, social consequences of actions and decisions rather than, or as well as, delivering earnest warnings about long-term repercussions.

10

When things go wrong

ALTHOUGH MOST ADOLESCENTS DEVELOP into mentally healthy adults, this period of life does confer vulnerability to mental illness. A survey of over nine thousand people in the United States in 2001–3 indicated that three-quarters of all cases of mental illness – including depression, anxiety, eating disorders, substance abuse and psychosis* – start at some point before the age of 24. Something about adolescence makes some people susceptible to psychiatric disturbance, especially if they have a genetic predisposition to mental illness and if they experience certain environmental risk factors.

Scientists are making progress in understanding what is happening in the brain development of adolescents who develop these conditions. There are new studies tracking brain maturation in children and adolescents over a long period of time, to see what happens to the structure of their brains if – and, more importantly, *before* – they develop a psychiatric condition.

There has been a substantial body of research on the emergence of schizophrenia, a devastating condition that affects around 1 per cent of the world's population. In the 1980s Irwin Feinberg suggested

* Psychosis describes a set of conditions in which the individual experiences a variety of disturbing symptoms, including hallucinations, delusions, confused and disturbed thoughts, and lack of insight and self-awareness about their illness. Examples of psychosis are schizophrenia, bipolar disorder and schizoaffective disorder.

that schizophrenia might be due to abnormal synaptic pruning. This idea was based on the knowledge gained from Huttenlocher's studies (described in chapter 4) that the human prefrontal cortex undergoes substantial synaptic pruning during adolescence. Feinberg speculated that synaptic pruning during adolescence might be exaggerated in young people who develop schizophrenia, so that their brains lose too many synapses during this critical period.

At the time, this was pure speculation, but recently several lines of evidence have provided support for this theory. Post-mortem studies have shown that the brain tissue of adults with schizophrenia contains fewer synapses than the brain tissue of people without schizophrenia. Brain-imaging studies have shown that grey matter volume in several brain regions, including the prefrontal cortex and the temporal cortex, is lower in young adults with schizophrenia than in individuals of the same age without this illness. However, while these findings are consistent with excessive synaptic pruning in the brains of adolescents who develop schizophrenia, they don't prove the connection, and other studies have shown different findings.

The key question is whether brain development unfolds differently in adolescents who develop schizophrenia. The only way to know this is to scan very large numbers of children and teenagers every year or so as they grow up. It is statistically likely that approximately 1 per cent of these young people will develop schizophrenia at some point in their late teens or early adulthood. It would then be possible to compare retrospectively brain development in the people who developed schizophrenia and in those who did not. However, to ensure a sufficient number of people in the schizophrenia group, you would need to scan tens of thousands of children over the course of their development, making this kind of study very challenging. It could be done, but it would take many years and use a huge amount of resources.*

* MRI scans are expensive. In the UK at the time of writing, they generally cost around £500 per hour, and each scan takes around one hour.

Another way of approaching this question is to study adolescents who are at high risk of developing schizophrenia, either because they have a parent with the illness (giving them a genetic predisposition) and/or because they already experience low levels of psychotic symptoms such as delusions and hearing voices. Studies including these high-risk groups have shown differences in grey and white matter development during adolescence in the young people who are later diagnosed with schizophrenia compared with those who are not. Specifically, the increase in white matter seen in typically developing adolescents is less pronounced in the high-risk adolescents who develop schizophrenia. These adolescents who go on to develop schizophrenia also show a greater decline in grey matter volume, particularly in the temporal cortex, than is seen in typically developing adolescents. In contrast, these deviations from the 'typical' trajectory are not seen in adolescents who are at high risk of schizophrenia but who do not develop the condition. This suggests that the trajectories of grey matter and white matter development are different in young people who develop schizophrenia – and importantly, that these divergences from the typical occur *before* they develop the illness.

One of the key messages from this research is that it is critically important that we look at brain *development* in mental disorders such as schizophrenia in order to try to understand how and why it occurs. Instead of taking just a snapshot of what the brain looks like in young adults with schizophrenia compared to young adults who don't have schizophrenia, we need to investigate *how* it came to look like this. It's just like working out the journey someone took in their car: you need to know the route, whether they took motorways or smaller roads, how congested the traffic was, and so on.

The same is true for cognitive abilities. A study published in 2006 showed that intelligence (as measured by IQ tests) is predicted not by brain structure *per se* but by the developmental trajectory of brain structure during adolescence. Philip Shaw and colleagues from the

NIMH in Bethesda carried out a longitudinal study scanning children and adolescents and measuring their IQ levels as they grew up. The results revealed that level of intelligence was associated with the pattern of brain development across adolescence, primarily in the frontal cortex. More intelligent children demonstrated a particularly rapidly changing cortex in late childhood and early adolescence, with an initial accelerated and prolonged phase of increase in cortical thickness, followed by rapid and substantial cortical thinning by early adolescence. The researchers speculated that the brains of more intelligent individuals are more 'plastic' (changeable) during this sensitive period of brain development.

While this result has yet to be replicated, a longitudinal study from 2014 of 504 participants aged 9–60 years also reported that developmental changes in cortical thickness were related to IQ. However, this study demonstrated that higher IQ was associated with cortical thinning in childhood and cortical thickening in adulthood. Although the results are slightly different from those of Shaw's study, both sets of findings underscore the importance of developmental changes in cortical thickness – that is, how much the cortex changes over time is more important to IQ than its eventual thickness. This highlights the more general point that the development of the brain has an impact on cognitive abilities in adulthood; the same is true for mental health.

∞

Other mental illnesses such as depression are also more likely to start in adolescence than at any other time of life. Of all mental health conditions that emerge during adolescence, depression is the one with the largest impact on health throughout the lifespan in terms of 'years lost to disability'.* There are several brain-imaging studies of adolescents

* The disability-adjusted life year (DALY) is a measure of overall disease burden, expressed as the number of years lost due to ill-health, disability or early death. It was developed in the 1990s as a way of comparing overall health and life expectancy in different countries.

with depression, pointing to different grey matter volumes in several cortical and subcortical regions compared to adolescents without depression. However, there is no clear consensus that one area of the brain is fundamentally different in adolescents with depression. Perhaps this is unsurprising – depression comes in many forms and has differing levels of severity, with many different symptoms and co-occurring conditions.

It appears to be more likely that it is the way the brain *functions*, and particularly the functional connections between brain regions, that is different in adolescents with depression. Neuroimaging studies of adolescents have shown increased levels of activity in regions that process emotion, such as the amygdala and the medial prefrontal cortex, when adolescents with depression carry out tasks with an emotional component, such as looking at emotional expressions in photographs of faces, or ignoring emotional faces, as compared with adolescents who do not have depression. There is also evidence that the connections between regions during emotion processing are stronger in adolescents with depression than in those without the condition. Some researchers have suggested that increased activity and connections within emotion-processing regions during tasks that involve emotional processing might reflect heightened reactivity to emotional and social stimuli in adolescents with depression.

One striking aspect of the various forms of mental illness is that they do not affect the genders equally. Depression, for example, is twice as common in women as in men. Eating disorders are also more common in women than in men, while men are more likely to suffer from substance- and alcohol-abuse disorders. Could these marked gender differences* provide a clue as to what is happening in the brain

* There's a difference between sex and gender. Sex is usually defined biologically, in terms of reproductive organs and sex chromosomes (in rare cases, this is difficult because some people are born with ambiguous sex organs, and others are born with an abnormal number of sex chromosomes). In contrast, gender is the state of being male or female, and is defined more by social and cultural differences, and how an individual feels, than by biological factors. In this book I use the term 'gender' rather than 'sex', because in most studies we simply ask people whether they are male or female; we don't check their biological sex by, for example, assessing their sex chromosomes.

to cause these illnesses?

One possible reason for the gender disparity is the difference in sex hormones between males and females (primarily testosterone in males and oestrogen in females), which start to diverge markedly at puberty. Indeed, the gender differences in the incidence of depression emerge at around mid-puberty, which may suggest that sex hormones contribute to the development of depression, and possibly other mental health problems that are characterized by gender differences. We know from many studies that hormones affect the brain.

Other factors might also play a role, including gender differences in brain development, and the different societal expectations placed on boys and girls. Overall, women are up to 40 per cent more likely than men to suffer from mental health problems, according to recent studies carried out in Western countries including the UK and the United States. Daniel Freeman, a psychiatrist at the University of Oxford, and his brother Jason Freeman, have written a book about the gender differences in mental illness, called *The stressed sex*. They argue that women tend to internalize their worries, which are then reflected in depression, self-harm and sleep problems, whereas men are more likely to externalize theirs, turning to alcohol and other substances. They also point to evidence that women's self-esteem is lower than men's, which might make women vulnerable to many mental health problems. However, women are more likely to talk about their symptoms than men – mental illness is still strongly stigmatized, perhaps especially so for men – and this might result in underdiagnosis of mental illness in men, and possibly overdiagnosis in women.

Some of the early brain-imaging studies reported striking gender differences in brain development. The early studies by Jay Giedd and his colleagues showed that grey matter volume in the prefrontal cortex peaks later in boys than in girls, at around the ages of 12 and 11, respectively. Because this gender difference in grey matter development occurs in early adolescence, it was attributed to gender differences in puberty onset: boys go through puberty around eighteen months to two years later than girls, on average. However, these early findings of gender differences in grey matter development have not been replicated by recent large-scale studies. Kate Mills and colleagues have found that the gender differences in brain development are largely explained by differences in head size and overall brain size, both of which are greater in boys than in girls. Controlling for these measurements in the analysis largely negated the gender differences.

So what are the causes of mental illness? This is the million-dollar question, and we are still very far from understanding why some people develop these conditions while others don't. There is a genetic component to most of them, meaning that someone with a parent who has a mental illness has a higher risk of developing that illness themselves, via their inherited genes. However, no mental illness is caused entirely by genetic factors: while some people have a genetic predisposition to a psychiatric disorder, certain environmental factors may elevate that risk – indeed, the condition may need environmental triggers to show itself at all. For example, research has shown that the chances of developing schizophrenia are higher for people who live in an urban rather than a rural environment.

The risk of developing schizophrenia is also increased in immigrants who live in a culture that is different from their own, and in which they are not always accepted by the society around them. Robin Murray, a professor of psychiatry at the Institute of Psychiatry in London, has studied the environmental risk factors for schizophrenia in young people in south London. One area of his research has focused on immigrants, especially those from African and Caribbean countries. His research has shown that there is an increased incidence of schizophrenia in ethnic minority immigrants, and suggests that social and cultural factors, such as racism and social isolation, play an important role.

Another environmental risk factor that Murray's research

highlighted is heavy cannabis use in adolescence. In the course of their studies, he and his colleagues interviewed thousands of young people about their cannabis use throughout their teens, and then kept in touch with these young people and took note of which of them developed schizophrenia. They found that prolonged and heavy use of cannabis during the teenage years is associated with an increased likelihood of developing schizophrenia in late adolescence or early adulthood. Using cannabis before the age of 15, using high-potency cannabis (such as skunk) and heavy use (four or five strong joints each day) each further increase the risk of developing schizophrenia.

One problem with this conclusion is that we know that cannabis use is higher in people with schizophrenia than in the general population. The question is, is the relationship causal, and if so, in which direction? Does cannabis cause schizophrenia, or is it the other way round: does having schizophrenia make you more likely to use cannabis? Some people have argued that perhaps young people with the early symptoms of schizophrenia, such as hearing voices and feeling paranoid, might choose to use cannabis to allay the fear aroused by these experiences – in other words, that cannabis is being used as a form of self-medication.

Disentangling cause and effect in this complex mix of factors is a challenge, but Murray has attempted to do just that. In his studies, children and young adolescents who reported potential early symptoms of schizophrenia at the start of his study were not included in the analysis, thereby reducing the likelihood that the results reflect an element of self-medication in the relationship between cannabis use and schizophrenia. This suggests that smoking cannabis in the teenage years can increase the risk of developing schizophrenia, at least in people with a genetic predisposition to the illness.

∞

People can develop schizophrenia without ever having tried cannabis. Most young people who smoke cannabis do not develop schizophrenia; this is even true for young people who smoke a large amount of skunk before the age of 15. But there are other detrimental effects of heavy cannabis use during the teenage years. A study published in 2012 by Madeline Meier, Avshalom Caspi, Terrie Moffitt and their colleagues provides some evidence that cannabis use in the teenage years might have a detrimental effect on cognitive ability in adulthood. The researchers investigated the association between cannabis use and cognitive ability in the 1,037 participants in the Dunedin longitudinal study (mentioned in chapter 9), who by the time of Meier's study were 38 years old.

This study looked at how persistent cannabis use – defined as using cannabis at least four days per week – at age 18 was associated with changes in IQ and various cognitive measures, including working memory, processing speed, perceptual reasoning and verbal comprehension, at age 38. Cognitive ability was tested at age 13, prior to first cannabis use, so the researchers had a baseline measure with which to compare the results of the second test in the same individuals 25 years later.

The findings showed that persistent cannabis use before age 18 was associated with a significant decline in cognitive ability, as measured by an IQ test, between childhood and adulthood. The researchers also found that the more persistent the cannabis use (by 'persistent', they meant regular use sustained over a long period), the greater the cognitive decline. Importantly, the association between cannabis use and cognitive decline was greater for those who began using cannabis before age 18 than for those who started using it later. If the onset of cannabis use was before the age of 18, the cognitive decline remained significant even if the individual had stopped using cannabis for at least one year before testing. This suggests that the cognitive effects weren't a consequence of current cannabis use, or of being stoned during testing. Instead, the negative effects of cannabis were still present long after the drug had been last taken.

The main take-home message from this study is that consuming cannabis before age 18 is more damaging for cognitive abilities than consuming it after that age. However, we still have to be cautious about interpreting the correlation between adolescent cannabis use and later cognitive ability as a direct causal relationship. Perhaps a third factor, for example decreased motivation, or a psychiatric condition developed in adolescence, such as anxiety or depression, leads people both to smoke cannabis in adolescence and to perform poorly on IQ and cognitive tests.

Nevertheless, this study is important because it suggests that adolescence might represent a sensitive period of development during which the environment can make its mark on how cognitive abilities develop. In this case the environmental factor was cannabis, which appears to have long-lasting negative consequences for cognitive abilities for some people, perhaps because adolescent cannabis use influences how brain circuitry develops.*

Whether the decline in cognitive ability associated with cannabis use before the age of 18 is reversible through training and rehabilitation programmes is a question for future research.

∞

What about alcohol? Does drinking influence brain development in adolescence? Many adolescents experiment with it – binge-drinking is common among young people in Western countries. Binge-drinking is usually defined as the consumption of four or more units per drinking occasion for females, and five or more units per drinking occasion for males. It's a subject that has attracted increasing attention from the media over recent years, owing to its direct association with rates of

* It's worth noting that not all teenagers who smoke cannabis are negatively affected by it. However, little is understood about what causes the cognitive capacities of some young people to be more affected than others.

risky behaviour, including behaviour that results in accidents and injuries. Over half of 15–18-year-olds in the UK and up to a third in the US report binge-drinking during the past month.

A small number of teenagers drink so much they are diagnosed with alcohol-use disorders, which are defined as problem drinking patterns that, over the course of one year, cause distress and affect their daily functioning, for example, interfering with school or work. It's important to say that most adolescents who have the occasional alcoholic drink, or even binge-drink every so often with their friends, do not progress to having alcohol-use disorders. The majority of young people who binge-drink in their teens or early twenties stop doing so, or do so much less frequently, once they assume adult roles and responsibilities such as working and building a family.

People tend to drink alcohol because of its immediate effects: it's a social lubricant and can make young people (and older people) feel more relaxed in a social group. But how does alcohol affect the developing brain? It is possible that alcohol use during the adolescent years has a higher potential for harming the brain than during adulthood.

Research in animals suggests that alcohol has negative consequences for the developing brain. Consumption of alcohol in rats during early adolescence triggers a chain of biological and behavioural changes, causing difficulties with, for example, movement and memory. Interestingly, adolescent animals seem to experience fewer negative effects from drinking alcohol than adult animals do, including hangover and sedative effects. This means that young adolescent animals might experience the positive aspects of alcohol without the negative consequences that deter adults from drinking more and more frequently. This might be why adolescent rats tend voluntarily to consume two or three times as much alcohol as adult rats if given the chance.

I became interested in how alcohol affects the human adolescent brain a few years ago, after meeting Sarah Feldstein-Ewing at a conference in the United States. Sarah is a clinical psychologist who works with young people with substance-abuse problems. At the time she was working in New Mexico, and she has since moved to Oregon. Both places have serious issues with drug and alcohol problems among young people – the television series *Breaking Bad* was based in New Mexico and, according to Sarah, was a pretty accurate portrayal of the drug problems there.

We decided to read all the published scientific papers looking at how alcohol affects the human adolescent brain and summarize the findings in a review. It turns out there's a large number of papers in this area – and also a fair amount of confusion, with some studies contradicting each other. The papers we were summarizing included brain-scanning studies of teenagers who were classified as having alcohol-abuse problems, or were routinely binge-drinking. One problem we faced is that not all of the papers used the same definitions of these two categories: some defined binge-drinking as four units in one sitting, others five units, others used a measure of drinks not units, and some of these didn't define what was meant by one drink – for example, a strong cocktail or a small beer? Another problem was that many of the studies examined only small numbers of participants. It's important to include a sufficiently large number of participants in brain-imaging studies if you're to be confident that your findings are valid.

Reading the papers, we found that only a handful of MRI studies had compared brain *structure* in adolescent drinkers and non-drinkers. Some of these studies found that adolescents who drink have smaller brain volumes in several regions, including parts of the prefrontal cortex that are involved in self-control and inhibition. The authors of these papers suggested that this might reflect difficulty in resisting the temptation to engage in risky but exciting and potentially rewarding activities, such as binge-drinking with friends. This would be a *cause* of drinking more. Alternatively, smaller brain volumes might be a *consequence* of drinking alcohol.

Some studies found that adolescent drinkers' brains showed smaller volumes in subcortical regions such as the hippocampus, which is involved in memory, and larger volumes in other regions. A small number of studies also showed poor white-matter connections between brain areas in adolescent drinkers. Some of the studies showed that, the more a young person drinks, the bigger the structural differences between their brain and those of non-drinking adolescents. This could suggest that the more a young person drinks, the worse the damage – but it might be the other way round: the bigger the difference in their brains, the more likely they are to drink. We simply cannot know from these correlational studies.

Many fMRI studies have measured brain activity during cognitive tasks in adolescent drinkers and non-drinkers. Various different tasks have been used, mostly related to memory, inhibition and gambling. A number of findings emerge from these studies.

First, while in almost all of these studies the drinkers were able to perform the tasks as well as the non-drinkers – their behaviour wasn't different – the drinkers nevertheless showed markedly different patterns of brain response. Many studies showed that adolescent drinkers showed *greater* engagement of regions involved in the task at hand, and also engaged numerous additional regions that aren't typically involved in these tasks. The authors of these studies suggested that this extra brain activity when completing a cognitive task might be due to compensation – that is, that the adolescent drinkers needed to activate more parts of their brain than the non-drinkers to achieve comparable performance.

∞

As I've already said more than once, the biggest problem with this kind of study is attributing cause and effect. We want to be able to investigate how drinking alcohol affects the developing brain. The best way to do this would be to recruit two groups of similar teenagers – similar socio-economic backgrounds, similar IQ scores and so on – and give one group a lot to drink over the course of their adolescence and the other group none. Of course, this isn't possible – it's a ludicrous idea and would be completely unethical. Instead, researchers study groups of young people who have already been diagnosed with an alcohol-abuse problem, and compare their brains with those of people of similar age who don't have such a problem.

There's a difficulty here too, though: these two groups are likely to differ in lots of ways, not just in how much alcohol they drink. The young people who have an alcohol-use problem are fairly likely to have co-occurring psychiatric problems such as depression or anxiety. They are also likely to abuse other drugs – it's rare that a teenager who abuses alcohol never uses any other substances. All of these additional factors are likely to affect the brain, so it's hard to attribute differences in their brains purely to their use of alcohol.

Almost all of the studies in this area are cross-sectional: that is, they involve a group of teenagers with alcohol-abuse problems being compared with a group of teenagers without such problems. Again, this means it is not possible to determine whether any brain differences between the two groups are *caused by* drinking alcohol or *predated* alcohol use. To establish that would require longitudinal studies, in which children are studied as they grow up, and the brains of those who develop alcohol-abuse problems are compared with the brains of those who don't. But there are very few longitudinal studies in this area. You can see why – fortunately, alcohol-abuse problems are not very common, so you would need to start with a very large number of children in order to end up with a group of young people who abuse alcohol.

As with all neuroimaging studies that compare groups, there is no clear way to interpret differences in brain structure and function between two groups. We are still far from understanding what higher volumes of grey matter, or higher levels of activation, in one group compared with another mean – are they 'good' or 'bad'? Higher levels of activation in the alcohol-abuse group could be interpreted as representing compensatory mechanisms, while lower activation may represent less recruitment of necessary brain systems.

Collectively, these studies suggest that young people who drink a lot show a different pattern of brain structure and function compared to those who do not. But there is room for optimism, as there is evidence to suggest that the adolescent brain may be able to get back on track once a young drinker becomes able to reduce or abstain from alcohol use. This is a reminder of how plastic and adaptable the adolescent brain is.

∞

What about other experiences during adolescence, such as video-gaming and using mobile phones – how do these affect the developing brain? This is an important question, and a lot has been written about the potentially damaging effects of screens on brain development. Many parents worry about the amount of time their children spend on phones, tablets and computers, and are sure that this must be harmful to their developing brains.* But is it?

A few years ago, Kate Mills carried out an analysis of the literature on the way screens affect the adolescent brain. Her conclusion was that nothing is known! The research simply has not been done – there are very few published scientific studies that have investigated the effect of playing video-games and surfing the internet on children's or adolescents' brains, and no longitudinal studies. We urgently need systematic and controlled studies looking at this important question.

* I am no exception. In my house, we have complicated rules about the ratio of screen-time to non-screen-time my children are allowed. They point out this rule doesn't seem to apply to me. This is an important observation, and there is as yet no research looking at how parental screen-time affects their children's development or their relationship with their children.

Perhaps screen-time is bad for children and adolescents, perhaps it isn't, or perhaps it depends on the amount of time they are spending on screens and/or what they are looking at.

It's important to remember that humans throughout the ages have always worried about the effects of new technologies on young people's minds. Plato seemed to be concerned that the invention of writing would harm people's memories:

> for this discovery of yours will create forgetfulness in the learners' souls, because they will not use their memories; they will trust to the external written characters and not remember of themselves. The specific which you have discovered is an aid not to memory, but to reminiscence, and you give your disciples not truth, but only the semblance of truth; they will be hearers of many things and will have learned nothing; they will appear to be omniscient and will generally know nothing; they will be tiresome company, having the show of wisdom without the reality.*

There were similar concerns about the effect of the printing press on young minds. In 1545 Conrad Gessner, a Swiss doctor, wrote that the abundance of written text would be 'confusing and harmful' to the mind. Since then there have been analogous concerns about the effects on our minds of the invention of television, the gramophone, radio and email.†

Contrary to popular belief about screens, a study in 2016 suggested that moderate amounts of screen-time are not detrimental to mental well-being. Andrew Przybylski of the University of Oxford and Netta

* We should be careful not to cherry-pick – in other passages, Plato was very positive about the invention of reading and writing. See https://senseandreference.wordpress.com/2010/10/27/reading-writing-and-what-plato-really-thought/.

† Vaughan Bell of UCL summarizes fears about modern technologies throughout the ages in this article: http://www.slate.com/articles/health_and_science/science/2010/02/dont_touch_that_dial.html.

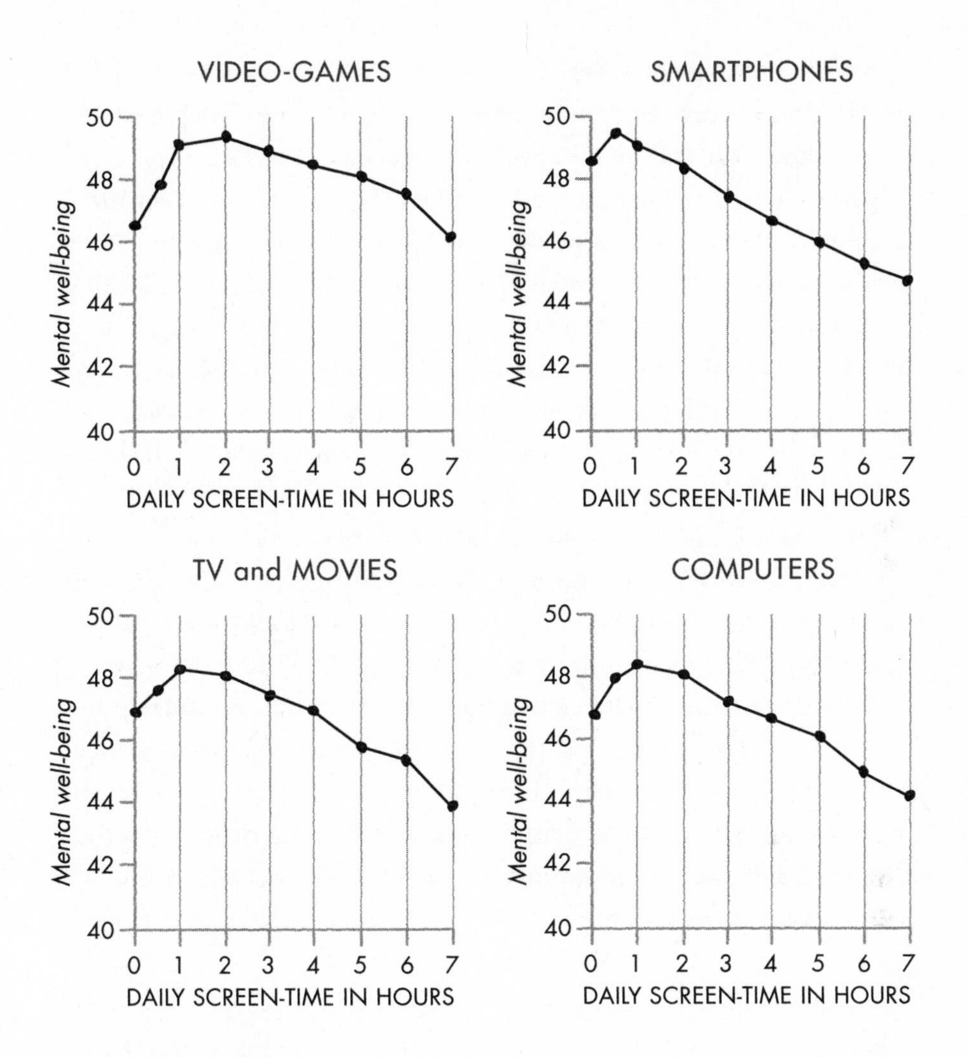

Well-being and screen-time among 15-year-olds

Weinstein of Cardiff University analysed reports on screen-time and mental health from over 120,000 15-year-olds in the UK registered on the Department for Education's National Pupil Database. Participants filled in a questionnaire about their mental health and answered questions about how much time they spent using different types of digital devices, including watching television, playing video-games, surfing the web, checking email and using smartphones for social activities.

The results revealed an 'inverted-U'-shaped relationship between screen-time and well-being. A very low level of screen-time was associated with low well-being, and well-being increased as screen-time increased, up to a certain point – after which, increased screen-time was associated with decreased well-being. On weekdays, well-being peaked in those who spent several hours on screens, including a mixture of video-gaming, smartphone use, watching videos and computer use – at both school and home, for school work, homework and leisure (see the illustration on the previous page). Often two or more screens were being used simultaneously, which is why the number of hours spent on screens seems high. Even above these 'tipping points', the link between screen-time and low well-being was fairly weak – in fact, it was less strong than the association between eating breakfast and well-being, or getting regular sleep and well-being.

We don't know where cause and effect lie in the relationship between screen-time and well-being – this study was correlational only. It's possible that low well-being causes young people to spend very little time – or a lot of time – on screens, rather than the other way round. It does, however, suggest that moderate screen-time is not detrimental to well-being.

∞

The studies described in this chapter suggest that environmental experiences, including use of drugs, alcohol and technology, may have impacts on adolescents' brains. The studies that have been carried out suggest that the brain might be particularly susceptible to environmental experiences during this period of life.

If the adolescent brain is particularly malleable, this might be a key time for the environment to exercise positive effects – and it is not too late to intervene with therapy in cases where individuals might need some extra help. It's critically important that we do not write off adolescents who are struggling with mental health issues. And yet, despite the fact that mental health problems in childhood and adolescence may be on the increase, in the UK only 6 per cent of the National Health Service's mental health budget, and 0.6 per cent of its overall budget, is dedicated to child and adolescent mental health services. This is an abysmal underinvestment in a very serious problem for individuals and for society. Recently, the British government has underlined the importance of understanding and treating mental illness in adolescence; this is a step in the right direction.

11

Educating the brain

In recent years, neuroscience has had an impact on our ideas about education. And there is no doubt: neuroscience is relevant to education. Every time you learn anything new, something in your brain – probably a few thousand synapses – changes. Education changes the brain, and therefore neuroscience is fundamental to teaching and learning.

The field of 'educational neuroscience' is flourishing, and has attracted the interest of many people, neuroscientists and educators alike. There are educational neuroscience journals, master's programmes, courses, books and so on. Since 2000, I've worked with education policy-makers and on education committees, and have written a book about neuroscience and education – *The learning brain* – with Uta Frith. In our book, Uta and I made the comparison between educators and gardeners. In the right soil, a healthy plant will grow. Just as plants can be nurtured through gardening, so the brain can be shaped and moulded through teaching.

However, using neuroscience to inform educational strategy isn't straightforward. In fact, the American philosopher John Bruer has argued that the gap between neuroscience findings and education practice presents a 'bridge too far' and that psychology studies, which investigate learning *behaviour*, are nearer and more relevant to schooling. He argues that neuroscience studies tend to be carried out in small samples, in laboratories, using controlled conditions and stimuli, and are a far cry from teaching and learning in busy classrooms.

It is also the case that there are as many myths about the brain as there are sound scientific findings that are genuinely relevant to education. The questions I'm asked when I give talks reveal a few that are particularly popular. One classic example is the enduring idea that people are 'left-brained' or 'right-brained'. It is sometimes suggested that our education system currently favours 'left-brain' modes of thinking, which are supposed to be logical, analytical and accurate, while not putting enough emphasis on right-brain modes of thinking, which are supposed to be creative, intuitive and emotional. The notion that the two sides (hemispheres) of the brain are involved in different 'modes of thinking' and that one hemisphere dominates the other in many individuals has become widespread, in particular in schools and workplaces. You can fill in questionnaires on Facebook to find out if you are left-brained or right-brained. There are websites that offer to teach you how to become more 'whole-brained'.

This idea is not based on knowledge of how the brain works! While it is true that the brain is made up of two hemispheres and one hemisphere is often initially active before the other during actions, speech and perception, in fact both sides of the brain work together in almost all situations, tasks and processes. The hemispheres are in constant communication with each other, and – with a few rare exceptions – it simply is not possible for one hemisphere to function without the other hemisphere 'joining in'. In other words, you are not right-brained or left-brained. You use both sides of the brain all the time, and you use them together.

Another popular notion is that we use just 10 per cent of our brains. The obvious question is then: what strategies should we develop better to employ the remaining 90 per cent? But there is no evidence for this 10 per cent idea. A large proportion of the brain is activated when you simply tap your finger. Tap your finger at the same time as you're reading this page and, given that you are also maintaining your heartbeat, breathing and body temperature, almost *all* of your brain will be active.

It's hard to pinpoint the origins of the 10 per cent myth. It may have come from the rare cases of individuals who had a brain scan that revealed most of their brain had never developed. Instead of the thick folds of cortex contained in most of our brains, they have only a very thin sliver of cortex. There have been one or two such cases of individuals who are also reported to have managed to live a normal life, going to school, holding down a job, and sustaining friendships and a family life. If it's possible to function to that extent with only a thin sheet of cortex, perhaps the rest of the brain is redundant?

In fact, cases like this demonstrate not that the rest of the cortex has no function, but that the brain is incredibly resilient: just a tiny proportion of brain cells can, in certain circumstances, enable a person to behave normally. This is especially the case if the person was *born* without part of their brain: through development, the brain may adapt to do its work with only a small proportion of the usual number of brain cells. This is a case of plasticity.

We should also bear in mind that people reported to be living completely normally with only a fraction of their brains may well not have undergone extensive cognitive assessments. So, superficially, they may seem to behave normally, but this might be because they have learned strategies to compensate for any difficulties caused by their condition. When tested on appropriately sensitive tasks, they might well show abnormalities. This is just one example of the need for rigorous scientific investigation to draw conclusions about the relationship between brain structure and brain function, and of the crucial need for data.

∞

Because research into the brain is of interest to the public, the results of a new study will often make a big splash in the media. But one big new experimental finding might be not replicated when the same scientist, or a different scientist, attempts to repeat it. This 'failure to replicate' is fairly common – and it's a controversial topic at the moment in the field of psychology.

A study published in 2016 attempted to replicate 100 well-known psychology studies, and reported that 70 per cent of the original findings could not be replicated. This may happen either because a finding is so small or sensitive to context that it is impossible to replicate, or because the finding simply isn't real. It is crucially important to understand that science doesn't produce facts, it produces findings – and these findings can be overturned by further research. Take the claim made twenty-five years ago that MMR vaccines cause autism. The results of one study were published in the highly respected medical journal *The Lancet*, and unsurprisingly gained huge traction. The 'discovery' made headline news. But the study couldn't be replicated: no other study, not even very large studies with thousands of participants, has ever found a link between vaccinations and autism. Meanwhile many children go unvaccinated.

The controversy around the 'Bruce Perry images' is another example of the need for proper data. During the 1990s, a psychiatrist based in Texas called Bruce Perry published a paper including images that he claimed were evidence of the detrimental consequences of severe neglect on the developing brain. One image was an MRI scan of a healthy 3-year-old child's brain; the other, an image of a very much smaller brain, purported to be a brain scan from a child of the same age who had been brought up in a Romanian orphanage. There is no question that the impact of these images is very powerful, and as a result they are frequently shown in education or policy contexts, often to make the point that early environments are crucial to brain development and should be invested in. But Perry gave no evidence of the case history of the neglected child; nor do we have any information about what a 'normal childhood' meant in this context.

Of course it is critically important that a child's early environment is warm and nurturing, and scientific studies have indeed shown that neglect is damaging to the developing brain. Two of my colleagues at UCL, Eamon McCrory and Essi Viding, have worked on this subject for many years and have shown that, in children and adolescents who have experienced neglect and/or abuse in early childhood, the brain develops differently in terms of both its structure and its activity during emotion-processing tasks.

However, the effects of neglect on the developing brain are more complicated and subtle than the Perry image suggests, and this image might be potentially misleading for several reasons. First, neglect can take many forms. The child depicted in this brain image might have experienced severe neglect associated with some institutional settings, possibly including sensory deprivation, other forms of abuse and malnutrition. Second, it is not possible to exclude other biological explanations for the differences between this image and that of the 'normal' 3-year-old's brain. Congenital neurological abnormalities would have to be ruled out before drawing any conclusion about the specific effect of neglect on the developing brain from these images. Third, it is not valid to make inferences about the impact of neglect on the basis of a single child. Studies with multiple participants and appropriate statistical analyses are required before meaningful conclusions about the general impact of neglect can be drawn. A large analysis of the field indicates that neglect and abuse are associated with specific and much more subtle differences in brain structure.

It might be argued that these images are simply illustrative of the potential effects of neglect and that it can't do any harm to show them. Perry himself suggests the neuroscience is persuasive:

> We've known about these outcomes [from early years trauma] well before we had any neuroscience to back it up . . . the role that neuroscience has played in this has been mostly to make

> people who are uncomfortable with social sciences think that somehow [the benefit of early intervention] is real.

To a certain extent I agree with this. If arresting and shocking images such as these alert people to the damaging consequences of neglect, that can only be a good thing. The problem is that images like this are potentially misleading, because the brains of most children who have suffered neglect do not show such substantial differences when viewed on MRI scans. But the fact that you can't usually detect differences from looking at a single brain image does not mean there are no effects of neglect on the brain. The use of these brain images out of context risks undermining understanding about the very real – but much more subtle – effects that neglect has on the developing brain.

The point Bruce Perry makes about the role of neuroscience is important. We trust the language of neuroscience and we want neuroscientific explanations for everything we do. It's not just neuroscience – we like scientific explanations to justify many of our choices. I'm as susceptible as anyone else – I appear to be seduced by shampoos that claim to contain 'neofibrine – a fusion of ceramides and an advanced shine-booster'. This sounds good to me, but the truth is I have no idea what any of that really means.

There is quite a lot of evidence that many of us find neuroscience evidence disproportionately persuasive. In one study, research participants were presented with both good and bad explanations of psychological phenomena, either with or without neuroscience terminology – and, critically, where the terminology was included it did not add anything substantive. The participants rated how satisfying and convincing they found the different explanations. Those new to neuroscience found the good explanation without the terminology more satisfying than the bad explanation without the terminology. However, with the addition of (unnecessary and unhelpful) neuroscience terms, they found the bad explanation more convincing.

This 'seductive allure of neuroscience' means that educationalists and policy-makers are likely to be particularly – and maybe unduly – influenced when considering education practice and interventions. There are also many products aimed at schools that refer to neuroscience in suggesting they can optimize children's brain development and learning. I'm afraid some of these make claims about the brain that have no basis in fact.

There are products used in many schools that purport to improve children's concentration and attention in lessons by getting children to do simple physical exercises before class. Some of these products make claims about improving the connections between the two brain hemispheres by doing coordination exercises: touching one elbow to the opposite knee, for example. Another exercise claims to improve oxygen flow to the brain by pressing 'brain buttons' that are found in the neck and behind the ribs.

This sort of claim makes no physiological sense. We don't have brain buttons on our bodies – they don't exist. Pressing your ribs or your neck will not increase oxygen flow to your prefrontal cortex; and even if it did, no one knows whether this would improve concentration. Moving your arm towards your opposite leg might involve the two brain hemispheres – but we don't know whether or how this affects attention or anything related to education.

Many teachers I speak to rate these products highly – they say doing these exercises does improve concentration and attention in the classroom, and that's what matters. This may be so, but the key point is that it has nothing necessarily to do with the claims being made about the brain. Perhaps doing a bit of exercise, having a break or doing something completely different is what causes improvements in concentration. Or maybe it's the placebo effect – the positive effect of an intervention that can result from the belief that it will work. In this case, the belief that an exercise will improve concentration may be enough to improve concentration for some people. Or perhaps these products might indeed change the brain – we just don't know, because the studies haven't been done.

Unfortunately, unlike medicine, which requires new drugs to undergo years of controlled testing in animals and humans before they are approved and licensed, education doesn't traditionally require any kind of trial for new products.

Isn't it important to know what aspect of a product is working, and how, before rolling it out into classrooms? Education changes children's brains, so we should care about what it consists of, in the same way we'd be careful about taking a drug that alters brain chemicals. In medicine, drugs are only approved if they have been through a randomized controlled trial – RCT for short. This is a rigorous and powerful way of finding out whether an intervention (a drug, for example) has a real effect on a symptom, or not. In an RCT, participants are allocated *at random* (by chance alone) to receive one of two or more interventions. One of the interventions is the drug under investigation, and one is a *control*. In clinical trials of drugs, the control might be a placebo 'sugar pill', which resembles the drug being tested but has no active ingredient. Participants should not know which group they're in – they shouldn't know whether the drug they are taking is the real drug or the sugar pill. This is called a *blind* RCT. A *double-blind* RCT is one in which neither the participant nor the scientists or doctors running the trial know which group is which. This helps minimize any bias the scientist might have when interacting with participants and when analysing whether one group has responded to the intervention more than the other group.

Ben Goldacre, a psychiatrist who runs the Evidence-Based Medicine Data Lab at the University of Oxford and writes prolifically about 'bad science', has written an interesting article on the importance of randomized controlled trials in schools. In it, he argues that education should be evidence-based in the same way that medicine is. Perhaps unsurprisingly, this has caused some controversy – some people don't think trials are necessary in education as they are in medicine. But I agree with Goldacre that it's important to test new methods of teaching systematically before putting them into practice in the classroom.

∞

One area of neuroscience that is relevant to education and that has undergone RCTs is brain training. Is it possible to train the brain? Of course it is! The brain is plastic, capable of change, and the brain changes whenever you learn something new. In this way, we are training our brains whenever new learning takes place. Brain training is big business – there are many apps and exercises on the market that claim to train your child's brain and increase their IQ (or indeed your own). Should you invest in a brain training app for your child? We know that children can be trained to complete tests more successfully, but will they become more intelligent? What is the science behind this?

Some studies have shown that training a cognitive skill not only improves the skill that has been trained, but also transfers to other cognitive skills. Working memory is the ability to remember information in your head for a short period of time, for example someone's phone number or the location of several objects on your desk. Torkel Klingberg and his colleagues in the Karolinska Institute in Stockholm, Sweden, have spent many years studying working memory training in children and adults. They have developed video-games that are designed to improve different types of working memory. In one game, you are in control of a rocket that flies around space, dismantling space mines by using your spatial working memory to remember where the mines were and in which order they appeared. Try it online – it's fun!*

In one of the first experiments led by Klingberg, published in 2002, children with attention-deficit hyperactivity disorder (ADHD) played

* http://www.spaceminespatrol.com/.

working memory video-games for several weeks. ADHD is a developmental condition that is associated with problems in working memory, and children with this condition might benefit particularly from working memory training. Before and after training with the video-games, the children completed tests of working memory and non-verbal reasoning – the ability to use logic and spot patterns to solve problems. After playing the video-games, participants performed better not only on tests of working memory but also on the reasoning tests. So this study suggested that working memory training not only improves working memory but also brings about improvements in other domains such as reasoning. This was an important finding because it held so much promise for children who are struggling with academic skills, including those with ADHD.

Further experiments showed similar effects in children without ADHD, and in adults, and demonstrated that working memory training also had benefits in other domains, including arithmetic. Other studies from other laboratories confirmed these results. Susanne Jaeggi and her colleagues at the University of California, Irvine, found that working memory training increased non-verbal reasoning capability, and also that the effect was dose-dependent: the more training participants did, the more their non-verbal reasoning improved.

As you might imagine, a flurry of brain-training products were created on the back of these exciting results. Schools and parents buy these products in the hope they will improve children's performance, and they are also popular with older adults who are trying to improve their memories. The problem is that many of the early studies were carried out in small samples and did not include a control group. A control group of participants trained on something other than working memory is important, because its inclusion is the only way to show that the training is specific to working memory and not just some general effect of the passage of time, or the placebo effect.

Better-controlled studies including large numbers of participants have largely failed to replicate the finding that working memory training improves anything except working memory. A study published in 2010 by Adrian Owen and colleagues at the Medical Research Council Cognitive and Brain Sciences Unit in Cambridge, UK, involved training 11,430 participants online over a six-week period on various different skills including working memory and reasoning. Performance improved on the skills that had been trained, but there was no improvement on non-trained tasks or in IQ.

Some studies, then, show that cognitive training does generalize to other skills, and sometimes also to IQ; other studies have not confirmed these results. This is still a controversial area, but an emerging pattern suggests that training does transfer to cognitive skills that are conceptually close to the trained skills. So, if you train on one type of memory task you'll be likely to improve on that memory task, and you might well get a bit better on tasks that test different types of memory; but it's unlikely that your general IQ will increase or that other skills, such as reasoning or language, will improve. If you want to get better at working memory, a working memory training app will probably help you. But it's unlikely to make you better at anything else.

Disagreements, findings and counter-findings are part and parcel of normal scientific progress and integral to the evolution of our understanding about the brain. Neuroscience is a relatively young area, and we need to be patient and not overinterpret what findings we have. The upshot is that whatever claims you read about neuroscience in the news should first be swallowed with a substantial swig of scepticism. Whenever you see a headline about a new result in neuroscience, the first two questions you should ask are: is the sample big enough, and has the result been replicated? If the answer to either or both is 'no', it doesn't necessarily mean that the claim should be disregarded – just that it should be viewed with an appropriate amount of caution.

∞

I have spent some time in this chapter addressing the problems of applying neuroscience to education, but there are areas where I believe recent findings are important. There are still many questions to be answered and a lot more work remains to be done, but, on the basis of the last twenty years of research, we can be pretty confident that, at the very least, adolescent brains are different from children's and adults' brains and they aren't static – they are in the process of development.

So, if childhood is seen as a major opportunity, or a sensitive period, for teaching, adolescence should be seen in just the same way. And yet it is only relatively recently that teenagers have been routinely educated in the West, and in many parts of the world a large proportion of teenagers still have no access to education. UNICEF estimates that 40 per cent of the world's teenagers today do not have access to secondary school.

As we have seen throughout this book, adolescence is a time of increased risk-taking. A certain amount of risk-taking is important: taking a risk can work out well, and even if it doesn't it can help us learn what works and what doesn't. We need to get used to making our own independent decisions, sometimes through trial and error, in adolescence. Risk-taking is useful in many contexts: in sports, trying new music and fashion, meeting new people, travelling to new places and so on. Risk-taking in an educational context, such as answering a question in class or making an educated guess in an exam question, is a vital skill that enables progress and creativity. Teenagers should be encouraged to take the right sort of risks.

We now know the adolescent brain is malleable and adaptable: research on brain development suggests that adolescence might be a period of relatively high neural plasticity, in particular in brain regions involved in decision-making, planning and social cognition. This might have implications for 'when to teach what', and could inform both curriculum design and teaching practice with the aim of ensuring that classroom activities exploit periods of neural plasticity that allow for maximal learning.

For example, there might be an optimal age at which to teach algebra and abstract reasoning, based on the natural changes in parts of the brain that are responsible for these activities. Before this age, perhaps those brain regions are simply not ready to perform such calculations and undertake abstract reasoning; afterwards, it might be a lot harder to learn these skills for the first time. We don't yet know whether this is the case, but the research I described in chapter 6 suggests that learning certain forms of non-verbal reasoning is more efficient in late adolescence than in early adolescence. Knowing whether there are brain-based age windows for learning will be useful when designing lesson plans and educational curricula.

∞

The environment can shape brain development in good and bad ways. How culture, alcohol, drugs, playing video-games, spending time on Instagram and Facebook, multi-tasking between instant messaging and homework, and so on, affect brain development is not well understood. The necessary studies have not been carried out. There might be negative effects on the brain from spending a lot of time on the internet every day. Alternatively, perhaps today's adolescents will be techno-savvy, super-efficient multi-taskers.

We do have studies that show that sleep deprivation has a detrimental effect on learning. When animals or humans learn a particular task – for example, the layout of a new environment or a set of complex rules – and are deprived of sleep the night following the training session, their performance on the task the following day is dramatically worse than it would have been if they had slept. This is because, during a night's sleep, the brain consolidates the learning that has occurred during the previous day. In studies investigating this process, volunteers are trained on a complex task during the day and their brain activity is recorded. That night, while asleep, the volunteers have their brains scanned again. These studies show that the same brain areas that were activated during the training become reactivated during sleep. The brain activity recorded during sleep is thought to reflect reinforcement of the learning that took place during the day. And participants' performance on the task improves the following day, after sleep, even though they haven't practised the task since the day before. The brain reactivations during sleep seem to be beneficial to memory and learning.

So far, so good. But survey after survey shows that teenagers are sleep-deprived (meaning they sleep for fewer than nine hours per night, which is the amount of sleep that most teenagers need). Why? One explanation, put forward by researchers including Russell Foster at the University of Oxford* and Mary Carskadon at Brown University in America, is that teenagers are made to get up for school at a time when they should still be asleep. This is because at puberty their circadian rhythms – the body clock – change so they are less sleepy in the evening and more sleepy in the morning.

Our circadian rhythms govern many bodily functions including body temperature, blood pressure and blood hormone levels. The circadian system also regulates our ability to be alert and think clearly. Both our physical ability and our mental alertness – including, for example, the ability to solve crossword puzzles – vary according to the time of day. Daylight is an important regulator of our circadian rhythms. The circadian clock resides in a part of the brain called the *suprachiasmatic nucleus* (SCN). The SCN regulates the synthesis of *melatonin* in the pineal gland during the night. Melatonin conveys to the body when it is night-time – in humans and some other animals, it causes feelings of sleepiness. The production of melatonin at the

* Watch Foster's brilliant TED talk about sleep: https://www.ted.com/talks/russell_foster_why_do_we_sleep.

'wrong' time of day is one of the main reasons why we feel so odd when travelling to different time zones – why we get 'jet-lag'.

After puberty, melatonin starts to be produced later at night, which is why many teenagers don't feel sleepy until much later in the evening. It is also the reason why they struggle to get up in the morning (unlike younger children, who are super-alert early in the morning and ready to sleep early in the evening). Forcing teenagers to go to school early in the morning, when according to their body clock they should still be asleep, is depriving them of sleep, and we know that sleep deprivation inhibits learning and lowers mood. Being forced to wake up early each morning during the week also results in what is called 'social jet-lag' as teenagers catch up on sleep at weekends, often sleeping until past midday. This is like going to a different time zone every weekend, and is probably as disorientating as jet-lag. Many researchers have suggested, therefore, that school start times should be later for teenagers, to be more in line with their circadian rhythms. This might both improve learning and motivation, and result in happier individuals.

∞

Another area of adolescent life in which neuroscience might have a positive impact is the juvenile justice system. Laurence Steinberg is a prominent advocate of the idea that findings about the adolescent brain should influence the legal treatment of teenagers, and he has been involved in many legal cases involving minors. In his book *Age of opportunity*, and several papers on the subject, Steinberg makes a clear case that minors should be treated differently from adults, first because their brains are still developing and have a greater capacity for change, and second because adolescents are more likely to act impulsively and take risks, especially with friends, than are adults.

In 2005, the US Supreme Court considered evidence about the adolescent brain when it was deciding whether to prevent judges from imposing the death sentence on minors, no matter how serious their crime. This was during the Roper *v.* Simmons case in which a 17-year-old boy had committed murder. Taking into account the new findings from neuroscience, the Supreme Court ruled that the death penalty for those who had committed crimes while under the age of 18 years was 'cruel and unusual punishment' and was therefore at odds with the US Constitution. Before 2005, it was possible to sentence minors to death; indeed, twenty-two people under 18 years were executed in the United States between 1976 and 2005.

The capacity for change in adolescence is large, arguably larger than in adulthood, perhaps because of the substantial changes that are taking place in the adolescent brain. This fits with the evidence that the majority of adolescents who engage in criminal behaviour don't continue to commit crimes as adults, as has been discussed at length by Terrie Moffitt. Many crimes committed by minors are carried out with friends, not when alone – just like a lot of adolescent risk-taking behaviour. Adolescents are more likely to make bad decisions when with their friends, as they are influenced by their peers. This tendency to go along with the peer group should be considered when assessing adolescent behaviour, even perhaps in the context of the criminal justice system. Of course, brain and behavioural development are not the only factors that should be considered when making legal decisions and new laws, but they should certainly be taken into account when discussing the legal treatment of adolescents.

12

It's the journey that matters

> Everything is a phase. Even the long one from 10 to 16 called adolescence or the Seventh Circle of Hell.
>
> Suzanne Moore, *The Guardian*, Sunday, 15 December 2013

Moody, violent, untrustworthy: adolescents are often stereotyped and maligned in alarming ways. But, as we have come to understand, adolescent-typical behaviour isn't mindless and destructive. It happens for a reason. Risk-taking, heightened self-consciousness and spending more time with friends are all symptoms of an important stage of brain development, signs of a journey that has to be made if adulthood is to be reached.

Until about twenty years ago, the unwelcome side of adolescent behaviour was put down to raging hormones and changes in schools and social life. We now know that the brain undergoes substantial development during adolescence, and this brain development probably contributes to the ways adolescents typically behave. This is a huge step forward, not only because it gives us a better insight into the changes young people go through and into their experience of what can be a difficult and turbulent period, but also because it opens a door into understanding why so many mental illnesses start in adolescence.

Learning about typical brain development is necessary if we are to identify how it differs in individuals who develop conditions such as depression, eating disorders and schizophrenia, all of which frequently become apparent for the first time during adolescence. In the future, being able to predict who is at risk of developing these mental illnesses on the basis of their neural developmental trajectories will be useful, because it will enable us to intervene earlier. Perhaps in the future we can learn from cases like my childhood friends Ben and Jon about when, precisely, an early intervention might have helped stave off the development of a disorder. We are still a long way from being able to do this, and research on the development of the brain in adolescence is still a young field. But the rapid expansion of studies in this area, together with a deeper understanding of the genetic and environmental factors involved in mental illness, gives us hope that this might be possible at some point.

It's easy to mock teenagers and collectively lament how badly behaved and useless they are. But many adolescents do amazing things – they are entrepreneurs, coders, musicians, athletes, artists and scientists. Adolescence is a time of heightened creativity and novel thinking, energy and passion. In my lab, when we're in the creative part of the scientific process, designing experiments or interpreting data, it's the youngest people in my group whose ideas I often seek. They may not have as much knowledge or experience as more senior colleagues, but they often have original points of view and provide a different angle on the problem.

Indeed, experimental studies have suggested that certain forms of creativity are highest in mid-adolescence. Eveline Crone at the University of Leiden has studied the development of creative thinking in adolescence. In psychology, creativity is often defined as the ability to generate ideas and solutions that are both original and feasible. Of course, in everyday life creativity is much more than this, but this definition has been useful in designing experimental tasks to measure it.

In one study, Crone and her colleagues looked at the development of divergent thinking using two different tasks. Divergent thinking involves solving a problem by generating a variety of possible solutions, and is often used in psychology studies as a way of measuring creativity. In the Alternate Uses Task, participants are asked to think of as many uses as possible for a common object – say, a brick. This might elicit answers such as a step, a paperweight, a diving aid and a dumbbell. This kind of task was used by Crone and colleagues to measure *verbal* divergent thinking.

To measure *visual* divergent thinking, the researchers used the Creativity Ability Test, which involves learning rules about the ways in which visually presented diagrams are similar and dissimilar. In the version of the task used by Crone and her colleagues, there were nine different diagrams, and participants were asked to select groups of three diagrams on the basis of similarity in some of their properties, which had to be different from those of the other six.

In both the verbal and visual tasks, participants were instructed to write down as many solutions as they could within ten minutes. Creativity was measured in three ways on both tasks: the researchers counted the number of solutions given in a limited time period (they called this *fluency*); the *flexibility* of responses, which is the generation of responses in different conceptual categories; and *originality*, which is the uniqueness of solutions and ideas compared with other people's responses.

Individuals aged 10–30 years were tested on these tasks, and the results were intriguing. Some aspects of creativity didn't change much during the age range tested, notably the *fluency* and *flexibility* of responses in the *verbal* divergent thinking task. However, adults produced more *original* answers than the adolescents on this task. In contrast, on the *visual* task mid-adolescents (aged 15–16 years) were more creative than both younger adolescents and adults.

A key aspect of divergent thinking tasks is that they don't have a correct solution. In addition to these divergent thinking tasks, participants in this study also completed *insight* tasks, which do have a correct solution and involve establishing associations between unrelated pieces of information and mentally restructuring the problem. Often participants find insight into the problem comes suddenly in an 'Aha!' experience, but they are usually unable to report the processes that led them to the solution.

Crone and her colleagues used two insight tasks in the visual domain: the Gestalt Completion Task, which involves seeing a whole picture from small fragments; and the Remote Associates Task, which involves seeing an image pop out from blurred images and patterns. Another task measured insight in the verbal domain, and involved finding a fourth related word when presented with three words. All three tasks involve restructuring and unifying complex or degraded information to find a single optimum solution. The results showed both visual and verbal creative insight continued to develop into late adolescence.

Crone's studies suggest that creativity is still developing in adolescence. Some aspects of creativity are still improving, and other components of creativity are actually *higher* in adolescence than at other times of life. Adolescents are creative, their brains are plastic and malleable and they are quick learners. Memories from our adolescence and early adulthood are more vivid and long-lasting than memories from any other time in our lives – this is called the 'reminiscence bump'. Adolescence is an exciting time, full of novel experiences, experimentation and exploration. It's not a period of life that should be neglected or maligned by society. The adolescents of today are our future, after all.

∞

Since 2013, I've had the privilege to work alongside a large group of exceptionally creative teenagers who are part of Company Three, a youth theatre company in London. The group, together with directors

Ned Glasier and Emily Lim, had seen my TED talk on the teenage brain,* as well as other videos about teenage development, and had been inspired to create a play around this theme.

Ned and Emily asked Kate Mills and me to visit the theatre group and discuss the science of the adolescent brain. Working with the group over many weekend and after-school workshops and drama sessions, we were able to help them create *Brainstorm*, their play about the teenage brain. This was an interpretation of the science of the teenage brain, written and performed entirely by teenagers. When I saw the first performance of *Brainstorm* at Platform in Islington, from the first scene onwards I was mesmerized by the imaginative interpretation of the research, and by the brilliant and moving performances of these talented young people. The play was not only innovative and clever, it was also incredibly poignant, telling the stories of the complex relationships between the young people and their parents, all set within the context of the developing adolescent brain. The play went on to be shown at several theatres in London, including two runs at the National Theatre, to critical acclaim, and was made into a television performance on the BBC.

It is important that we find new ways to communicate our scientific discoveries to young people and the general public, and *Brainstorm* is a great example of this. The impact of the play on its audiences has been profound and long-lasting. It's an incredibly powerful drama, full of emotions. The cast told us stories of parents rethinking how they understand and interact with their children as a consequence of learning about brain development from the play. We heard about headteachers who saw the play and returned to their schools determined to do things differently. One young actor in the play described it as a 'journey of self-discovery'.

And, as scientists, we learned from the experience too. It's

* https://www.ted.com/talks/sarah_jayne_blakemore_the_mysterious_workings_of_the_adolescent_brain.

fascinating and important to learn about how the science of the adolescent brain is interpreted by young people themselves, to learn about their experiences, what's important to them and what they care about, and to use these insights to form ideas for our next studies.

Here is a short extract from the play, in which the teenagers are imagining their parents talking to them about their brains:

SAMA — You say to me

YAAMIN — Your brain is broken.
It's like an adult's brain, but it doesn't work properly.

SERAFINA — You say

YAAMIN — When you become a teenager something happens.
Your brain shrinks or something.
It stops working properly.
It gets so full of thinking about yourself that you forget about anyone else.

KASSIUS — You say

YAAMIN — That's why you're like this right now.
That's why you just did that.
That's why you just said that.

MICHAEL — You say

YAAMIN — Your brain is messed up.
It's like you're in some city you've never been to and you don't have a map and you don't know what you're doing.
And you keep taking the wrong turns.
You say
Listen to me.

One day you'll be ok.
Probably.
Your brain will start working properly.
One day your brain will be just like mine
And then you'll be ok.
But until then
You gotta try and be more . . . like me.

SEGEN I look at you.
I don't say anything.

SAMA I pick up my plate.

JACK Put it in the kitchen.

NOAH And go upstairs.

At the end of the play, the teenagers respond to their parents:

YAAMIN I say to you,
My brain isn't broken.
It's beautiful.
I'm in a city I've never been to and I see bright lights and new ideas and fear and opportunity and a thousand million roads all lit up and flashing.
I say
There are so many places to explore but you've forgotten that they exist because every day you walk the same way with your hands in your pockets and your eyes on the floor.
I say
If my brain changed overnight
If it just happened like that
Then all I'd have seen is what you have shown me
And all I'd be is a rubbish version of you.
I don't want to be a rubbish version of you

I want to be an amazing version of me
I say
When I'm wild and out of control
It's because I'm finding out who I am
What the world is
All the things I might be
And if I was a real wild animal
Then I'd have left by now.
But I haven't.
And I'm not going to. Yet.
I say
My brain isn't broken
It's like this for a reason
I'm becoming who I am
And I'm scared
And you're scared
Because who I am might not be who you want me to be.
Or who you are
And I don't know why, but I *don't* say
It'll all be ok.
There are so many things I don't say to you anymore.
So many things I write down but I can't say.
I want to say them
But I can't.
I pick up my plate
Put it in the kitchen
And go upstairs.

∞

As I have argued, adolescence is the period of life in which we develop a profound sense of who we are, and particularly of how we are seen by other people. The long journey of adolescence equips us with a sense of self-identity and an understanding of other people that enable us to become independent adults, no longer so reliant on our parents and families, and more established within our peer groups. Adolescence is a time of development and change, just as childhood is.

When, at the beginning of the twentieth century, Stanley Hall coined the term 'adolescence' to describe youth between childhood and adulthood, he considered it a stage of development based on changing experience. Some of the characteristics that defined adolescents for Hall included an increased engagement in risky behaviours and a strong dependence on friendships. What he could not have known is that both of these, and much else, are founded on substantial and protracted changes in the brain. Rather than being a period of purely social change, adolescence should be considered a unique stage of biological and psychological development. It's remarkable that Hall's estimation of the duration of adolescence, as up to age 25 years, corresponds pretty closely with the age at which brain development starts to level off.

Hall famously described adolescence as a period of 'storm and stress', characterized by uncontrollable and conflicting emotions. Hall talked about the egoism and vanity of this period of life, when energy and exuberance are followed by indifference, lethargy and melancholy. He saw adolescent behaviour as reflecting a combination of uninhibited childish selfishness and an increasing idealism and virtue. Although many of Hall's stereotypes about this age group have been dismissed as caricatures, he was right to highlight the importance of friendships and peer influence, as well as the urge to explore the world with curiosity and energy and to experience novelty. These are characteristics that are the focus of much of today's research on the adolescent brain.

Individual differences are lost in stereotypes. As I have argued throughout this book, there's no such thing as an average adolescent, and brain development varies widely between people. Research is starting to investigate what gives rise to these large individual differences – the person's genes, their specific environment – and the consequences of the individual brain's developmental trajectory across adolescence.

The teenage brain isn't broken. Adolescence is a period of life when the brain is changing in important ways: we should understand it, nurture it – and celebrate it.

Notes

1: Adolescence isn't an aberration

p. 2, 'ending between 22 and 25 years': G. S. Hall, *Adolescence*, 2 vols (New York: Appleton, 1916).

p. 2, 'stable, independent role in society': G. C. Patton, S. M. Sawyer, J. S. Santelli, D. A. Ross, R. Afifi et al., 'Our future: a *Lancet* commission on adolescent health and wellbeing', *Lancet*, vol. 387, 2016, pp. 2423–78.

p. 3, 'A study led by Laurence Steinberg': L. Steinberg, G. Icenogle, E. P. Shulman, K. Breiner, J. Chein, D. Bacchini, L. Chang, N. Chaudhary, L. D. Giunta, K. A. Dodge, K. A. Fanti, J. E. Lansford, P. S. Malone, P. Oburu, C. Pastorelli, A. T. Skinner, E. Sorbring, S. Tapanya, L. M. Tirado, L. P. Alampay, S. M. Al-Hassan and H. M. Takash, 'Around the world, adolescence is a time of heightened sensation seeking and immature self-regulation', *Developmental Science*, Jan. 2017 (online), doi: 10.1111/desc.12532.

p. 4, 'during the month or so of adolescence': S. Macrì, W. Adriani, F. Chiarotti and G. Laviola, 'Risk taking during exploration of a plus-maze is greater in adolescent than in juvenile or adult mice', *Animal Behaviour*, vol. 64, no. 4, 2002, pp. 541–6.

p. 4, 'A study published in 2014': S. Logue, J. Chein, T. Gould, E. Holliday and L. Steinberg, 'Adolescent mice, unlike adults, consume more alcohol in the presence of peers than alone', *Developmental Science*, vol. 17, no. 1, 2014, pp. 79–85.

p. 5, 'They just destroy everything': Calla Wahlquist, 'Woman attacked by wombat thought she was going to die', *Guardian*, 22 Aug. 2016, https://www.theguardian.com/australia-news/2016/aug/22/woman-attacked-by-wombat-thought-she-was-going-to-die.

p. 5, 'Aristotle described "youth"': Aristotle, *Rhetoric*, bk 2, ch. 12.

p. 6, '"I would there were no age"': William Shakespeare, *The Winter's Tale* (1611), Act III, scene iii.

p. 6, '"A change in humour"': Rousseau, *Emile, or On Education* (1762).

p. 10n., '*Schizophrenia genesis*': Irving I. Gottesman, *Schizophrenia genesis: the origins of madness* (Basingstoke and New York: W. H. Freeman, 1990).

p. 11, 'Chris Frith and Daniel Wolpert and I found': C. D. Frith, S.-J. Blakemore and D. M. Wolpert, 'Abnormalities in the awareness and control of action', *Philosophical Transactions: Biological Sciences*, vol. 255, 2000, pp. 1771–88.

2: A sense of self

p. 19, 'a letter sent by Dinah Hall of Lustleigh, Devon': Dinah Hall, 'Brief letters', *Guardian*, 4 Jan. 2013, http://www.theguardian.com/theguardian/2013/jan/04/pickles-in-a-sweat.

p. 20, 'some simple tests with newborn babies': P. Rochat, 'Five levels of self-awareness as they unfold early in life', *Consciousness and Cognition*, vol. 12, no. 4, 2003, pp. 717–31; P. Rochat (ed.), *The self in infancy: theory and research*, Advances in Psychology, no. 112 (Amsterdam: North Holland, Elsevier Science, 1995).

p. 21, 'than at a video of themselves': L. E. Bahrick and L. Moss, 'Development of visual self-recognition in infancy', *Ecological Psychology*, vol. 8, no. 3, 1996, pp. 189–208.

p. 21, 'the mirror self-recognition test': D. J. Povinelli, 'The unduplicated self', in Rochat (ed.), *The self in infancy*, pp. 161–92.

p. 21, 'differentiate between themselves and other people in speech': E. Bates, 'Language about me and you: pronominal reference and the emerging concept of self', in D. Cicchetti and M. Beeghly (eds), *The self in transition: infancy to childhood* (Chicago: University of Chicago Press, 1990), pp. 165–82.

p. 24, 'sometimes called the "looking-glass self"': Charles Cooley, *Human nature and the social order* (New York: Scribners, 1902). See https://en.wikipedia.org/wiki/Looking_glass_self.

p. 24, 'know us for different qualities, for example': Roy F. Baumeister and Brad J. Bushman, 'The self', in *Social psychology and human nature*, 2nd edn (Belmont, CA: Cengage Learning, 2011), pp. 57–96.

p. 24, 'the opinions of others': J. G. Parker et al., 'Peer relationships, child development, and adjustment: a developmental psychopathology perspective', in D. Cicchetti and D. J. Cohen (eds), *Developmental psychopathology*, vol. 1: *Theory and methods*, 2nd edn (New York: Wiley, 2006), pp. 419–93; L. R. Vartanian, 'Revisiting the imaginary audience and personal fable constructs of adolescent egocentrism: a conceptual review', *Adolescence*, vol. 35, no. 140, 2000, pp. 639–61.

p. 24, 'a striking effect of embarrassment': L. H. Somerville, 'The teenage brain:

sensitivity to social evaluation', *Current Directions in Psychological Science*, vol. 22, no. 2, 2013, pp. 121–7.

p. 25, 'may overestimate the extent to which this actually occurs': D. K. Lapsley and M. N. Murphy, 'Another look at the theoretical assumptions of adolescent egocentrism', *Developmental Review*, vol. 5, no. 3, 1985, pp. 201–17.

p. 25, 'The term "imaginary audience"': D. Elkind, 'Egocentrism in adolescence', *Child Development*, vol. 38, no. 4, 1967, pp. 1025–34.

p. 26, 'remains quite high even in adulthood': K. Frankenberger, 'Adolescent egocentrism: a comparison among adolescents and adults', *Journal of Adolescence*, vol. 23, no. 3, 2000, pp. 343–54.

p. 27, 'Leonora Weil investigated this question': Leonora G. Weil et al., 'The development of metacognitive ability in adolescence', *Consciousness and Cognition*, vol. 22, no. 1, 2013, pp. 264–71.

p. 28, 'With Suparna Choudhury, who was a PhD student at the time': S.-J. Blakemore et al., 'Adolescent development of the neural circuitry for thinking about intentions', *Social Cognitive and Affective Neuroscience*, vol. 2, no. 2, 2007, pp. 130–9.

3: Fitting in

p. 32, 'They designed a driving video-game': M. Gardner and L. Steinberg, 'Peer influence on risk taking, risk preference, and risky decision making in adolescence and adulthood: an experimental study', *Developmental Psychology*, vol. 41, no. 4, 2005, pp. 625–35.

p. 34, 'people aged 16–25 have more car accidents': Insurance Institute for Highway Safety, *Fatality facts: teenagers 2013* (Arlington, VA: Insurance Institute for Highway Safety, Highway Loss Data Institute, 2013), http://www.iihs.org/iihs/topics/t/teenagers/fatalityfacts/teenagers; http://www-nrd.nhtsa.dot.gov/Pubs/812021.pdf; L. Chen, S. P. Baker, E. R. Braver and G. Li, 'Carrying passengers as a risk factor for crashes fatal to 16- and 17-year-old drivers', *Journal of the American Medical Association*, vol. 283, no. 12, 2000, pp. 1578–82; Brake: The Road Safety Charity, *Young drivers* (2016), http://www.brake.org.uk/too-young-to-die/15-facts-a-resources/facts/488-young-drivers-the-hard-facts (see also http://www.brake.org.uk/assets/docs/dl_reports/DLreport6-Youngdrivers2011-full.pdf.

p. 34, 'special schemes in which "black boxes" are fitted': E. Walsh and V. Barford, 'The proliferation of the little black box', BBC News, 31 Oct. 2012, http://www.bbc.co.uk/news/magazine-20143969.

p. 34, 'Kate Mills, Anne-Lise Goddings and I': S.-J. Blakemore and K. L. Mills, 'Is adolescence a sensitive period for socio-cultural processing?', *Annual Review of*

Psychology, vol. 65, 2014, pp. 187–207; K. L. Mills, A.-L. Goddings and S.-J. Blakemore, 'Drama in the teenage brain', *Frontiers for Young Minds*, vol. 2, no. 16, 2014, pp. 1–5.

p. 34–5, 'In one series of studies published in the 1990s': R. Larson and M. H. Richards, 'Daily companionship in late childhood and early adolescence: changing developmental contexts', *Child Development*, vol. 62, no. 2, 1991, pp. 284–300.

p. 35, 'spending more and more time with their friends': Ibid.

p. 35, 'More recent surveys have found marked cultural differences': R. W. Larson and S. Verma, 'How children and adolescents spend time across the world: work, play and developmental opportunities', *Psychological Bulletin*, vol. 125, no. 6, 1999, pp. 701–36.

p. 36, 'the opinions of their peers become more important': R. W. Larson, M. H. Richards, G. Moneta, G. Holmbeck and E. Duckett, 'Changes in adolescents' daily interactions with their families from ages 10 to 18: disengagement and transformation'. *Developmental Psychology*, vol. 32, no. 4, 1996, pp. 744–54, doi:10.1037/0012-1649.32.4.744.

p. 36, 'When interviewed about friendships': S. F. O'Brien and K. L. Bierman, 'Conceptions and perceived influence of peer groups: interviews with preadolescents and adolescents', *Child Development*, vol. 59, no. 5, 1988, pp. 1360–5.

p. 36, 'adolescents feel a particular concern about being socially excluded': Blakemore and Mills, 'Is adolescence a sensitive period for socio-cultural processing?'.

p. 37, 'A few years ago, Catherine Sebastian and I': C. Sebastian et al., 'Social brain development and the affective consequences of ostracism in adolescence', *Brain and Cognition*, vol. 72, no. 1, 2010, pp. 134–45.

p. 37, 'originally devised over twenty years ago': K. D. Williams, 'Social ostracism', in R. M. Kowalski (ed.), *Aversive interpersonal behaviors* (New York: Plenum, 1997), pp. 133–70; K. D. Williams, 'Ostracism', *Annual Review of Psychology*, vol. 58, 2007, pp. 425–52.

p. 38, 'affects the way the brain develops in adolescent rats': M. P. Leussis and S. L. Andersen, 'Is adolescence a sensitive period for depression? Behavioural and neuroanatomical findings from a social stress model', *Synapse*, vol. 62, no. 1, 2008, pp. 22–30.

p. 38, 'Male adolescent rats exposed to social instability': M. R. Green, B. Barnes and C. M. McCormick, 'Social instability stress in adolescence increases anxiety and reduces social interactions in adulthood in male Long-Evans rats', *Developmental Psychobiology*, vol. 55, no. 8, 2012, pp. 849–59.

p. 38, 'changes in behaviour and hormone production': A. R. Burke, C. M.

McCormick, S. M. Pellis and J. L. Lukkes, 'Impact of adolescent social experiences on behavior and neural circuits implicated in mental illnesses', *Neuroscience and Biobehavioral Reviews*, 2017, pii: S0149-7634(16)30050-1; doi: 10.1016/j.neubiorev.2017.01.018 (online publication ahead of print).

pp. 38–9, 'Compared with rats that had stable social environments': C. M. McCormick et al., 'Deficits in male sexual behavior in adulthood after social instability stress in adolescence in rats', *Hormones and Behaviour*, vol. 63, no. 1, 2013, pp. 5–12.

p. 39, 'Adolescents in very socially unstable environments': T. Polihronakis, *Information packet: mental health care issues of children and youth in foster care* (New York: National Resource Center for Family-Centered Practice and Permanency Planning, 2008), http://www.hunter.cuny.edu/socwork/nrcfcpp/downloads/information_packets/Mental_Health.pdf.

p. 40, 'sometimes even tended to overestimate the risks': V. Reyna and F. Farley, 'Risk and rationality in adolescent decision making', *Psychological Science in the Public Interest*, vol. 7, no. 1, 2006, pp. 1–44; D. Romer, V. F. Reyna and T. D. Satterthwaite, 'Beyond stereotypes of adolescent risk taking: placing the adolescent brain in developmental context', *Developmental Cognitive Neuroscience*, vol. 27, 2017, pp. 19–34.

p. 40, 'risky decisions in so-called "hot" contexts': S.-J. Blakemore and T. W. Robbins, 'Decision-making in the adolescent brain', *Nature Neuroscience*, vol. 15, no. 9, 2012, pp. 1184–91.

p. 41, 'Kate Mills, Anne-Lise Goddings and I proposed a "see-saw" model': K. L. Mills, A.-L. Goddings and S.-J. Blakemore, 'Drama in the teenage brain', *Frontiers for Young Minds*, vol. 2, no. 16, 2014, pp. 1–5.

p. 43, 'many people are doing the same thing': Charles McKay, *Extraordinary popular delusions and the madness of crowds* (London: Richard Bentley, 1841).

p. 44, 'Even though adolescents understand': S.-J. Blakemore, 'Avoiding social risk in adolescence', *Current Directions in Psychological Science*, in press.

p. 44, 'One study that began in 1965': M. Obschonka, H. Andersson, R. K. Silbereisen and M. Sverke, 'Rule-breaking, crime, and entrepreneurship: a replication and extension study with 37-year longitudinal data', *Journal of Vocational Behaviour*, vol. 83, no. 3, 2013, pp. 386–96.

p. 45, 'an experiment to look at social influence on risk perception': L. Knoll et al., 'Social influence on risk perception during adolescence', *Psychological Science*, vol. 26, no. 5, 2015, pp. 583–92.

p. 46, 'We recently replicated these effects': L. J. Knoll, J. T. Leung, L. Foulkes and S.-J. Blakemore, 'Age-related differences in social influence on risk perception depend on the direction of influence', *Journal of Adolescence*, vol. 60, 2017, pp. 53–63.

p. 47, 'A 2016 study, carried out by researchers at Yale and Princeton universities': E. L. Paluck, H. Shepherd and P. M. Aronow, 'Changing climates of conflict: a social network experiment in 56 schools', *Proceedings of the National Academy of Sciences of the United States*, vol. 113, no. 3, 19 Jan. 2016, pp. 566–71.

p. 49, '"cultural niches" during adolescence': S. Choudhury, 'Culturing the adolescent brain: what can neuroscience learn from anthropology?', *Social Cognitive and Affective Neuroscience*, vol. 5, nos 2–3, 2010, pp. 159–67.

4: Inside the skull

p. 54, 'the brain contains around 86 billion nerve cells': F. A. Azevedo, L. R. Carvalho, L. T. Grinberg, J. M. Farfel, R. E. Ferretti, R. E. Leite, W. Jacob Filho, R. Lent and S. Herculano-Houzel, 'Equal numbers of neuronal and nonneuronal cells make the human brain an isometrically scaled-up primate brain', *Journal of Comparative Neurology*, vol. 513, no. 5, April 2009, pp. 532–41, doi: 10.1002/cne.21974.

p. 57, 'that of Phineas Gage': J. Fleischman, *Phineas Gage: a gruesome but true story about brain science* (New York: Houghton Mifflin, 2002).

p. 58, 'he was "no longer Gage"': J. M. Harlow, 'Recovery from the passage of an iron bar through the head', *Publications of the Massachusetts Medical Society*, vol. 2, no. 3, 1868, pp. 327–47 (repr. David Clapp & Son, 1869).

p. 59, '"He had all of these pictures of synapses in our house"': W. Yardley, 'Peter Huttenlocher, explorer of the brain, dies at 82', *New York Times*, 26 Aug. 2013, http://www.nytimes.com/2013/08/27/us/peter-huttenlocher-explorer-of-the-brains-development-dies-at-82.html?_r=0.

p. 60, '"more interesting than the abnormal population"': N. Stafford, 'Peter Huttenlocher', Obituaries, *British Medical Journal*, 22 Oct. 2013, p. 347, http://www.bmj.com/content/347/bmj.f6136.

p. 62, 'Janet Werker, Renée Desjardins and their colleagues in Canada': J. F. Werker and R. N. Desjardins, 'Listening to speech in the 1st year of life: experiential influences on phoneme perception', *Current Directions in Psychological Science*, vol. 4, no. 3, 1995, pp. 76–81.

p. 62, 'This result was discovered by Annette Karmiloff-Smith': M. Rivera-Gaxiola, G. Csibra, M. H. Johnson and A. Karmiloff-Smith, 'Electrophysiological correlates of cross-linguistic speech perception in native English speakers', *Behavioural Brain Research*, vol. 111, nos 1–2, 2000, pp. 13–23.

p. 63, 'a study by Patricia Kuhl': P. K. Kuhl, F. M. Tsao and H. M. Liu, 'Foreign-language experience in infancy: effects of short-term exposure and social interaction on phonetic learning', *Proceedings of the National Academy of Sciences of the United States*, vol. 100, no. 15, 22 July 2003, pp. 9096–101.

p. 63, 'levelling out at around 12 years': J. Easton, 'Peter Huttenlocher, pediatric neurologist, 1931–2013', *UChicagoNews*, 19 Aug. 2013, http://news.uchicago.edu/article/2013/08/19/peter-huttenlocher-pediatric-neurologist-1931-2013.

p. 66, 'by Oliver Sacks': O. Sacks, *The man who mistook his wife for a hat* (London: Duckworth, 1985).

p. 66, 'more recently, by Paul Broks': P. Broks, *Into the silent land: travels in neuropsychology* (London: Atlantic, 2003).

p. 66, 'that of Henry Gustav Molaison': S. Kean, *The tale of the dueling neurosurgeons: the history of the human brain as revealed by true stories of trauma, madness and recovery* (London: Transworld, 2014).

p. 66, 'the first paper about his poor memory': W. B. Scoville and B. Milner, 'Loss of recent memory after bilateral hippocampal lesions', *Journal of Neurology, Neurosurgery and Psychiatry*, vol. 20, no. 11, 1957, pp. 11–21.

5: Inside the living brain

p. 71, 'MRI uses a magnetic field': S. C. Bushong and G. Clarke, *Magnetic resonance imaging: physical and biological principles*, 4th edn (St Louis, Mo.: Mosby, 2015).

p. 73, 'Steve Fleming, had already carried out a study': S. M. Fleming, R. S. Weil, Z. Nagy, R. J. Dolan and G. Rees, 'Relating introspective accuracy to individual differences in brain structure', *Science*, vol. 329, no. 5998, 17 Sept. 2010, pp. 1541–3, doi: 10.1126/science.1191883.

p. 74, 'Eleanor Maguire and her colleagues at the Wellcome Trust Centre': E. A. Maguire, D. G. Gadian, I. S. Johnsrude, C. D. Good, J. Ashburner, R. S. Frackowiak and C. D. Frith, 'Navigation-related structural change in the hippocampi of taxi drivers', *Proceedings of the National Academy of Sciences of the United States*, vol. 97, no. 8, 11 April 2000, pp. 4398–403.

p. 75, 'Ochsner and his colleagues found': K. N. Ochsner, S. A. Bunge, J. J. Gross and J. D. Gabrieli, 'Rethinking feelings: an FMRI study of the cognitive regulation of emotion', *Journal of Cognitive Neuroscience*, vol. 14, no. 8, 2002, pp. 1215–29.

p. 76, 'One of the first developmental neuroimaging studies on the self': J. H. Pfeifer, M. D. Lieberman and M. Dapretto, '"I know you are but what am I?!": neural bases of self- and social knowledge retrieval in children and adults', *Journal of Cognitive Neuroscience*, vol. 19, no. 8, Aug. 2007, pp. 1323–37.

6: The ever-plastic brain

p. 79, 'It is one of the later stages of maturation': Gordon M. Shepherd, *Neurobiology*, 2nd edn (Oxford: Oxford University Press, 1988), p. 195.

p. 80, 'Giedd began a very large longitudinal study': J. N. Giedd, J. Blumenthal, N. O. Jeffries, F. X. Castellanos, H. Liu, A. Zijdenbos and J. L. Rapoport, 'Brain development during childhood and adolescence: a longitudinal MRI study', *Nature Neuroscience*, vol. 2, no. 10, 1999, pp. 861–3.

p. 80, 'A 2016 study by Christian Tamnes and Kate Mills': K. L. Mills, A. L. Goddings, M. M. Herting, R. Meuwese, S.-J. Blakemore, E. A. Crone and C. K. Tamnes, 'Structural brain development between childhood and adulthood: convergence across four longitudinal samples', *NeuroImage*, vol. 141, 2016, pp. 273–81.

p. 80, 'The analysis revealed': C. K. Tamnes, M. M. Herting, A. L. Goddings, R. Meuwese, S.-J. Blakemore, R. E. Dahl, B. Güroğlu, A. Raznahan, E. R. Sowell, E. A. Crone and K. L. Mills, 'Development of the cerebral cortex across adolescence: a multisample study of interrelated longitudinal changes in cortical volume, surface area and thickness', *Journal of Neuroscience*, vol. 27, no. 12, 2017, pp. 3402–12.

p. 84, 'Reactions to events in the outside world can be faster': J. I. Stiles and T. L. Jernigan, 'The basics of brain development', *Neuropsychology Review*, vol. 20, no. 4, Dec. 2010, pp. 327–48.

p. 85, 'This could be due to an increase': T. Paus, M. Keshavan and J. N. Giedd, 'Why do many psychiatric disorders emerge during adolescence?', *Nature Reviews Neuroscience*, vol. 9, no. 12, 2008, pp. 947–57.

p. 85, 'White matter volume stops increasing': C. Lebel, M. Gee, R. Camicioli, M. Wieler, W. Martin and C. Beaulieu, 'Diffusion tensor imaging of white matter tract evolution over the lifespan', *NeuroImage*, vol. 60, 2012, pp. 340–52.

p. 85, 'MRI studies have shown': Tamnes et al., 'Development of the cerebral cortex across adolescence'; Mills et al., 'Structural brain development between childhood and adulthood'.

p. 86, 'observed under the microscope': R. J. Zatorre, R. D. Fields and H. Johansen-Berg, 'Plasticity in gray and white: neuroimaging changes in brain structure during learning', *Nature Neuroscience*, vol. 15, no. 4, 2012, pp. 528–36.

p. 86, 'twice as many synapses as an adult brain': P. R. Huttenlocher and A. S. Dabholkar, 'Regional differences in synaptogenesis in human cerebral cortex', *Journal of Comparative Neurology*, vol. 387, no. 2, 1997, pp. 167–78; K. S. Mix, J. Huttenlocher and S. C. Levine, *Quantitative development in infancy and early childhood*, Infant Child Development, vol. 12 (Oxford: Oxford University Press, 2002).

p. 86, 'Zdravko Petanjek and his collaborators': Z. Petanjek et al., 'Extraordinary neoteny of synaptic spines in the human prefrontal cortex', *Proceedings of the National Academy of Sciences of the United States*, vol. 108, no. 32, 9 Aug. 2011, pp. 13281–6.

p. 88, 'the "Shopping Task"': T. Shallice and P. W. Burgess, 'Deficits in strategy application following frontal lobe damage in man', *Brain*, vol. 114, no. 2, 1991, pp. 727–41.

p. 88, 'Patients with injury to this region': https://www.headway.org.uk/about-brain-injury/individuals/effects-of-brain-injury/executive-dysfunction/.

p. 90, 'One of the first studies to assess the development of inhibition': B. Luna, K. Velanova and C. F. Geier, 'Development of eye-movement control', *Brain and Cognition*, vol. 68, no. 3, 2008, pp. 293–308.

p. 91, 'sensory input from the environment during development': M. H. Johnson, 'Interactive specialization: a domain-general framework for human functional brain development?', *Developmental Cognitive Neuroscience*, vol. 1, no. 1, Jan. 2011, pp. 7–21, doi: 10.1016/j.dcn.2010.07.003.

p. 91, 'This study was carried out by Lisa Knoll': L. J. Knoll, D. Fuhrmann, A. Sakhardande, F. Stamp, M. Speekenbrink and S.-J. Blakemore, 'A window of opportunity for cognitive training in adolescence', *Psychological Science*, vol. 27, no. 12, 2016, pp. 1620–31.

p. 94, '"Brain plasticity and the ability to change behavior"': Center on the Developing Child at Harvard University, *A decade of science informing policy: the story of the National Scientific Council on the Developing Child*, 2014, http://developingchild.harvard.edu/wp-content/uploads/2015/09/A-Decade-of-Science-Informing-Policy.pdf.

p. 94, 'in terms of the money saved': J. J. Heckman, 'Skill formation and the economics of investing in disadvantaged children', *Science*, vol. 312, no. 5782, 2006, pp. 1900–02, http://doi.org/10.1126/science.1128898.

p. 95, 'The part of the brain that processes sound': C. Pantev, R. Oostenveld, A. Engelien, B. Ross, L. E. Roberts and M. Hoke, 'Increased auditory cortical representation in musicians', *Nature*, vol. 392, no. 6678, 1998, pp. 811–14.

p. 95, 'The degree of enlargement is correlated': P. Schneider, M. Scherg, H. G. Dosch, H. J. Specht, A. Gutschalk and A. Rupp, 'Morphology of Heschl's gyrus reflects enhanced activation in the auditory cortex of musicians', *Nature Neuroscience*, vol. 5, no. 7, 2002, pp. 688–94.

p. 95, 'adults who were new to the piano': A. Pascual-Leone, D. Nguyet, L. G. Cohen, J. P. Brasil-Neto, A. Cammarota and M. Hallett , 'Modulation of muscle responses evoked by transcranial magnetic stimulation during the acquisition of new fine motor skills', *Journal of Neurophysiology*, vol. 74, no. 3, 1995, pp. 1037–45.

p. 96, 'before and after they had practised juggling': B. Draganski, C. Gaser, V. Busch, G. Schuierer, U. Bogdahn and A. May, 'Neuroplasticity: changes in grey matter induced by training', *Nature*, vol. 427, no. 6972, 2004, pp. 311–12.

7: Social mind, social brain

p. 97, 'the *social brain*': R. Adolphs, 'The social brain: neural basis of social knowledge', *Annual Review of Psychology*, vol. 60, 2009, pp. 693–716; C. D. Frith, 'The social brain?', *Philosophical Transactions of the Royal Society B*, vol. 362, no. 1480, 2007, pp. 671–8.

p. 99, 'the *mind-blindness* hypothesis of autism': Uta Frith, *Autism: explaining the enigma*, pb edn (Oxford: Blackwell, 2003).

p. 100, 'babies aren't able to tell us explicitly': A. Slater and P. C. Quinn, 'Face recognition in the newborn infant', *Infant and Child Development*, vol. 10, nos 1–2, 2001, pp. 21–4.

p. 101, 'Johnson and Morton suggested': M. Johnson and J. Morton, *Biology and cognitive development: the case of face recognition* (Oxford: Blackwell, 1995), pp. 9–10.

p. 101, 'to which they were exposed in the womb': C. Moon, H. Lagercrantz and P. K. Kuhl, 'Language experienced in utero affects vowel perception after birth: a two-country study', *Acta Paediatrica*, vol. 102, no. 2, 2013, pp. 156–60.

p. 101, 'preferring to listen to her speech': A. R. Webb, H. T. Heller, C. B. Benson and A. Lahav, 'Mother's voice and heartbeat sounds elicit auditory plasticity in the human brain before full gestation', *Proceedings of the National Academy of Sciences of the United States*, vol. 112, no. 10, 2015, pp. 3152–7.

p. 102, 'In a second study, Carey and Diamond': S. Carey, R. Diamond and B. Woods, 'Development of face recognition: a maturational component?' *Developmental Psychology*, vol. 16, no. 4, 1980, pp. 257–69.

p. 102, 'Another behavioural study': R. F. McGivern, J. Andersen, D. Byrd, K. L. Mutter and J. Reilly, 'Cognitive efficiency on a match to sample task decreases at the onset of puberty in children', *Brain and Cognition*, vol. 50, no. 1, 2002, pp. 73–89.

p. 103, 'development of subcortical brain regions': A.-L. Goddings, K. L. Mills, L. S. Clasen, J. N. Giedd, R. M. Viner and S.-J. Blakemore, 'The influence of puberty on subcortical brain development', *NeuroImage*, vol. 88, 2014, pp. 242–51.

p. 105, 'a non-linear pattern of development across age': I. Dumontheil, R. Houlton, K. Christoff and S.-J. Blakemore, 'Development of relational reasoning during adolescence', *Developmental Science*, vol. 13, 2010, pp. F15–F24.

p. 105, 'This pattern of performance was mirrored': Ibid.

p. 107, 'devised some clever experiments': A. Gopnik, 'Children can teach us that if we really want to learn about the world, we need to open ourselves up to new possibilities', *Psychology: Science in Action*, American Psychological Society, 2014, http://www.apa.org/action/careers/improve-lives/alison-gopnik.aspx.

p. 108, 'the *Sally–Anne Task*': S. Baron-Cohen, A. M. Leslie and U. Frith, 'Does the autistic child have a "theory of mind"?', *Cognition*, vol. 21, no. 1, 1985, pp. 37–46.

p. 108, 'problems in forming social relationships': U. Frith, *Autism and Asperger syndrome* (Cambridge: Cambridge University Press, 1991).

p. 108, 'false-belief understanding is present in much younger children': R. Baillargeon, M. Rose, R. M. Scott and Z. He, 'False-belief understanding in infants', *Trends in Cognitive Science*, vol. 14, no. 3, 2010, pp. 110–18.

p. 109, 'Baillargeon and Kristine Onishi measured the eye movements of babies': K. H. Onishi and R. Baillargeon, 'Do 15-month-olds understand false beliefs?', *Science*, vol. 308, no. 5719, 2005, 255–8.

p. 109, 'Another study, carried out at Birkbeck': A. Senju, V. Southgate, C. Snape, M. Leonard and G. Csibra, 'Do 18-month-olds really attribute mental states to others? A critical test', *Psychological Sciences*, vol. 22, no. 7, 2011, pp. 878–80.

p. 110, 'they often show the experimenter where the toy is hidden': Ibid.

p. 110, 'A further series of studies by Michael Tomasello': D. Buttelmann, M. Carpenter and M. Tomasello, 'Eighteen-month-old infants show false belief understanding in an active helping paradigm', *Cognition*, vol. 112, no. 2, 2009, pp. 337–42.

p. 111, 'if administered after mid-childhood': I. Dumontheil, I. A. Apperly and S.-J. Blakemore, 'Online usage of theory of mind continues to develop in late adolescence', *Developmental Science*, vol. 13, no. 2, 2010, pp. 331–8.

p. 111, 'the *Director Task*': B. Keysar, D. J. Barr, J. A. Balin and J. S. Brauner, 'Taking perspective in conversation: the role of mutual knowledge in comprehension', *Psychological Science*, vol. 11, no. 1, 2000, pp. 32–8.

p. 113, 'a computerized version of the Director Task': I. Dumontheil et al., 'Online usage of theory of mind continues to develop'.

p. 115, 'A study carried out by Berna Güroğlu and Eveline Crone': B. Güroğlu, W. van den Bos and E. A. Crone, 'Fairness considerations: increasing understanding of intentionality during adolescence', *Journal of Experimental Child Psychology*, vol. 104, no. 4, 2009, pp. 398–409.

p. 116, 'Anne-Kathrin Fett at the University of Amsterdam': A. K. J. Fett et al., 'Trust and social reciprocity in adolescence – a matter of perspective taking', *Journal of Adolescence*, vol. 37, no. 2, 2014, pp. 175–84.

p. 117, 'Findings from Wouter van den Bos': B. Güroğlu, W. van den Bos and E. A. Crone, 'Sharing and giving across adolescence: an experimental study examining the development of prosocial behavior', *Frontiers in Psychology*, vol. 5, no. 291, 2014, pp. 1–13.

8: Understanding other people

p. 119, 'The first brain-scanning study of mentalizing': P. C. Fletcher et al., 'Other minds in the brain: a functional imaging study of "theory of mind" in story comprehension', *Cognition*, vol. 57, no. 2, 1995, pp. 109–28.

p. 121, 'in which participants looked at cartoons': H. L. Gallagher, F. Happé, N. Brunswick, P. C. Fletcher, U. Frith and C. D. Frith, 'Reading the mind in cartoons and stories: an fMRI study of "Theory of Mind" in verbal and nonverbal tasks', *Neuropsychologia*, vol. 38, no. 1, 2000, pp. 11–21.

p. 122, 'In total we analysed 857 brain scans': K. L. Mills, F. Lalonde, L. S. Clasen, J. N. Giedd and S.-J. Blakemore, 'Developmental changes in the structure of the social brain in late childhood and adolescence', *Social Cognitive and Affective Neuroscience*, vol. 9, no. 1, 2014, pp. 123–31.

p. 123, 'A brain-imaging study carried out by Mirella Dapretto's research team': A. T. Wang, S. S. Lee, M. Sigman and M. Dapretto, 'Developmental changes in the neural basis of interpreting communicative intent', *Social Cognitive and Affective Neuroscience*, vol. 1, no. 2, 2006, pp. 107–21.

p. 124, 'in which we investigated social emotions': S. Burnett, G. Bird, J. Moll, C. Frith and S.-J. Blakemore, 'Development during adolescence of the neural processing of social emotion', *Journal of Cognitive Neuroscience*, vol. 21, no. 9, 2009, pp. 1736–50.

p. 127, 'a social cognition task and a working memory task': K. L. Mills, I. Dumontheil, M. Speekenbrink and S.-J. Blakemore, 'Multitasking during social interactions in adolescence and early adulthood', *Royal Society Open Science*, vol. 2, no. 11, 2015, doi:10.1098/rsos.150117.

p. 129, '"I had been working with my assistant"': O. Sacks, 'Face-blind: why are some of us terrible at recognizing faces?', *New Yorker*, 30 Aug. 2010, http://www.newyorker.com/magazine/2010/08/30/face-blind.

p. 129, 'car and bird experts were scanned': I. Gauthier, P. Skudlarski, J. C. Gore and A. W. Anderson, 'Expertise for cars and birds recruits brain areas involved in face recognition', *Nature Neuroscience*, vol. 3, no. 2, 2000, pp. 191–7.

p. 130, 'better at remembering the identity of faces': G. Golarai et al., 'Differential development of high-level visual cortex correlates with category-specific recognition memory', *Nature Neuroscience*, vol. 10, no. 4, 2007, pp. 512–22.

p. 130, 'increasingly selective for faces with age': S. K. Scherf, M. Behrmann, K. Humphreys and B. Luna, 'Visual category-selectivity for faces, places and objects emerges along different developmental trajectories', *Developmental Science*, vol. 10, no. 4, 2007, pp. 15–30.

p. 130, 'slightly different regions for each task': K. Cohen Kadosh, M. H. Johnson,

F. Dick, R. Cohen Kadosh and S.-J. Blakemore, 'Effects of age, task performance, and structural brain development on face processing', *Cerebral Cortex*, vol. 23, no. 7, 2013, pp. 1630–42.

p. 130, 'In a different study by Christopher Monk': C. S. Monk et al., 'Adolescent immaturity in attention-related brain engagement to emotional facial expressions', *NeuroImage*, vol. 20, no. 1, 2003, pp. 420–8.

9: The right sort of risks

p. 133, 'the leading cause of death in adolescence': R. M. Viner, C. Coffey, C. Mathers, P. Bloem, A. Costello, J. Santelli and G. C. Patton, '50-year mortality trends in children and young people: a study of 50 low-income, middle-income, and high-income countries', *Lancet*, vol. 377, no. 9772, 2011, pp. 1162–74.

p. 134, 'risk-taking in adolescence can lead to injury and illness': G. C. Patton, S. M. Sawyer, J. S. Santelli, D. A. Ross, R. Afifi et al., 'Our future: a *Lancet* commission on adolescent health and wellbeing', *Lancet*, vol. 387, 2016, pp. 2423–78.

p. 134, 'survival rates of North American high-school students': T. Willoughby, R. Tavernier, C. Hamza, P. J. Adachi and M. Good, 'The triadic systems model perspective and adolescent risk taking', *Brain and Cognition*, vol. 89, Aug. 2014, pp. 114–15.

p. 134, 'his excellent book': L. D. Steinberg, *Age of opportunity: lessons from the new science of adolescence* (New York: Houghton Mifflin, 2014).

p. 135, '"the dual systems model"': L. Steinberg, 'A dual systems model of adolescent risk-taking', *Developmental Psychobiology*, vol. 52, no. 3, 2010, pp. 216–24.

p. 136, 'carried out an analysis in collaboration': K. L. Mills, A.-L. Goddings, L. S. Clasen, J. N. Giedd and S.-J. Blakemore, 'The developmental mismatch in structural brain maturation during adolescence', *Developmental Neuroscience*, vol. 36, nos 3–4, 2014, pp. 147–60.

p. 136n., 'in a 2016 paper': L. H. Somerville, 'Searching for signatures of brain maturity: what are we searching for?', *Neuron*, vol. 92, no. 6, 2016, pp. 1164–7.

p. 141, 'One of the earliest studies of this kind': M. Ernst, E. E. Nelson, S. Jazbec, E. B. McClure, C. S. Monk, E. Leibenluft, J. Blair and D. S. Pine, 'Amygdala and nucleus accumbens in responses to receipt and omission of gains in adults and adolescents', *NeuroImage*, vol. 25, no. 4, 2005, pp. 1279–91.

p. 142, 'Subsequent fMRI studies': A. Galvan, T. A. Hare, C. E. Parra, J. Penn, H. Voss, G. Glover and B. J. Casey, 'Earlier development of the accumbens relative to orbitofrontal cortex might underlie risk-taking behavior in adolescents', *Journal of Neuroscience*, vol. 26, no. 25, 2006, pp. 6885–92; S.-J. Blakemore and T. W. Robbins, 'Decision-making in the adolescent brain', *Nature Neuroscience*, vol. 15, no. 9, 2012, pp. 1184–91.

p. 142, 'When participants chose high-risk gambles': L. Van Leijenhorst, B. Gunther Moor, Z. A. Op de Macks, S. A. Rombouts, P. M. Westenberg and E. A. Crone, 'Adolescent risky decision-making: neurocognitive development of reward and control regions', *NeuroImage*, vol. 51, no. 1, 2010, pp. 345–55.

p. 143n., 'other animals, including rats': T. L. Doremus-Fitzwater, E. L. Varlinskaya and L. P. Spear, 'Motivational systems in adolescence: possible implications for age differences in substance abuse and other risk-taking behaviors', *Brain and Cognition*, vol. 72, no. 1, 2010, pp. 114–23; R. M. Philpot and L. Wecker, 'Dependence of adolescent novelty-seeking behavior on response phenotype and effects of apparatus scaling', *Behavioral Neuroscience*, vol. 122, no. 4, 2008, pp. 861–75.

p. 143n., 'a larger proportion of neurons in the dorsal striatum': D. A. Sturman and B. Moghaddam, 'Striatum processes reward differently in adolescents versus adults', *Proceedings of the National Academy of Sciences of the United States*, vol. 109, no. 5, 2012, pp. 1719–24.

p. 143, 'when decisions are made in an emotional – "hot" – context': L. H. Somerville, R. M. Jones and B. J. Casey, 'A time of change: behavioral and neural correlates of adolescent sensitivity to appetitive and aversive environmental cues', *Brain and Cognition* 72: 1, 2010, pp. 124–33.

p. 144, 'In a study by Casey, Somerville and Todd Hare': L. H. Somerville, T. Hare and B. J. Casey, 'Frontostriatal maturation predicts cognitive control failure to appetitive cues in adolescents', *Journal of Cognitive Neuroscience*, vol. 23, no. 9, 2011, pp. 2123–34.

p. 145, 'the ability to think counterfactually': A. A. Baird and J. A. Fugelsang, 'The emergence of consequential thought: evidence from neuroscience', *Philosophical Transactions of the Royal Society B: Biological Sciences*, vol. 359, no. 1451, 2004, pp. 1797–1804.

p. 145, 'the development of counterfactual emotions in making decisions': S. Burnett, N. Bault, G. Coricelli and S.-J. Blakemore, 'Adolescents' heightened risk-seeking in a probabilistic gambling task', *Cognitive Development*, vol. 25, no. 2, 2010, pp. 183–96.

p. 147, 'a version of the Iowa Gambling Task': E. Cauffman, E. P. Shulman, L. Steinberg, E. Claus, M. T. Banich, S. Graham and J. Woolard, 'Age differences in affective decision making as indexed by performance on the Iowa Gambling Task', *Developmental Psychology*, vol. 46, no. 1, 2010, pp. 193–207.

p. 147, 'still developing in adolescence': B. J. Casey, R. M. Jones and T. A. Hare, 'The adolescent brain', *Annals of the New York Academy of Sciences*, vol. 1124, 2008, pp. 111–26.

p. 148, 'in "cold" tasks, with no emotional context': E. A. Crone, L. Bullens, E. A.

van der Plas, E. J. Kijkuit and P. D. Zelazo, 'Developmental changes and individual differences in risk and perspective taking in adolescence', *Development and Psychopathology*, vol. 20, no. 4, 2008, pp. 1213–29; D. J. Paulsen, M. L. Platt, S. A. Huettel and E. M. Brannon, 'Decision-making under risk in children, adolescents, and young adults', *Frontiers in Psychology*, vol. 2, 2011, article 72.

p. 148, 'As impulse control gradually improves': A. Scheres, M. Dijkstra, E. Ainslie, J. Balkan, B. Reynolds, E. Sonuga-Barke and F. X. Castellanos, 'Temporal and probabilistic discounting of rewards in children and adolescents: effects of age and ADHD symptoms', *Neuropsychologia*, vol. 44, no. 11, 2006, pp. 2092–2103.

p. 148, 'Brain-scanning studies have shown': A. Christakou, M. Brammer and K. Rubia, 'Maturation of limbic corticostriatal activation and connectivity associated with developmental changes in temporal discounting', *NeuroImage*, vol. 54, no. 2, 2011, pp. 1344–54.

p. 150, 'able to concentrate and to exert self-control': W. Mischel, Y. Shoda and P. K. Peake, 'The nature of adolescent competencies predicted by preschool delay of gratification', *Journal of Personality and Social Psychology*, vol. 54, no. 4, 1988, pp. 687–96; Y. Shoda, W. Mischel and P. K. Peake, 'Predicting adolescent cognitive and social competence from preschool delay of gratification: identifying diagnostic conditions', *Developmental Psychology*, vol. 26, no. 6, 1990, pp. 978–86.

p. 150, 'the better they did in their school exams': W. S. Mischel, Y. Shoda and M. I. Rodriguez, 'Delay of gratification in children', *Science*, vol. 244, no. 4907, 1989, pp. 933–8.

p. 150, 'quicker at responding to the "go" stimuli': I. M. Eigsti, V. Zayas, W. Mischel, Y. Shoda, O. Ayduk, M. B. Dadlani, M. C. Davidson, J. Lawrence Aber and B. J. Casey, 'Predicting cognitive control from preschool to late adolescence and young adulthood', *Psychological Science*, vol. 17, no. 6, 2006, pp. 478–84.

pp. 150–1, 'This study was led by B. J. Casey': B. J. Casey, L. H. Somerville, I. H. Gotlib, O. Ayduk, N. T. Franklin, M. K. Askren, J. Jonides, M. G. Berman, N. L. Wilson, T. Teslovich, G. Glover, V. Zayas, W. Mischel and Y. Shoda, 'Behavioral and neural correlates of delay of gratification 40 years later', *Proceedings of the National Academy of Sciences of the United States*, vol. 108, no. 36, Sept. 2011, pp. 14998–15003.

p. 151n., 'Mischel has written a wonderful book about his research': W. Mischel, *The marshmallow test: mastering self-control* (London: Transworld, 2014).

p. 152, 'composite measure of childhood self-control': T. E. Moffitt, L. Arseneault, D. Belsky, N. Dickson, R. J. Hancox, H. Harrington, R. Houts, R. Poulton, B. W. Roberts, S. Ross, M. R. Sears, W. M. Thomson and A. Caspi, 'A gradient of childhood self-control predicts health, wealth, and public safety', *Proceedings of the National Academy of Sciences of the United States*, vol. 108, no. 7, 2011, pp. 2693–8.

p. 153, 'Mindfulness is a state of being': Mark Williams and Danny Penman, *Mindfulness: a practical guide to finding peace in a frantic world* (London: Piatkus, 2011).

p. 153, 'Our study, led by Stefano Palminteri': S. Palminteri, E. J. Kilford, G. Coricelli and S.-J. Blakemore, 'The computational development of reinforcement learning during adolescence', *PLoS Computational Biology*, vol. 12, no. 6, 2016, e1004953.

p. 154, 'impact of the following different messages': C. Pechmann and E. G. Reibling, 'Anti-smoking advertising campaigns targeting youth: case studies from USA and Canada', *Tobacco Control*, vol. 9, no. 2, 2000, pp. i18–i31.

p. 155, 'manipulate them into unhealthy eating': C. J. Bryan, D. S. Yeager, C. P. Hinojosa, A. Chabot, H. Bergen, M. Kawamura and F. Steubing, 'Harnessing adolescent values to motivate healthier eating', *Proceedings of the National Academy of Sciences of the United States*, vol. 113, no. 39, 2016, pp. 10830–5, http://www.pnas.org/content/113/39/10830.abstract.

10: When things go wrong

p. 157, 'start at some point before the age of 24': R. C. Kessler, P. Berglund, O. Demler, R. Jin, K. R. Merikangas and E. E. Walters, 'Lifetime prevalence and age-of-onset distributions of DSM-IV disorders in the National Comorbidity Survey Replication', *Archives of General Psychiatry*, vol. 62, no. 6, 2005, pp. 593–602.

p. 157, 'certain environmental risk factors': T. Paus, M. Keshavan and J. N. Giedd, 'Why do many psychiatric disorders emerge during adolescence?' *Nature Reviews Neuroscience*, vol. 9, no. 12, 2008, pp. 947–57.

p. 158, 'lose too many synapses during this critical period': I. Feinberg, 'Schizophrenia: caused by a fault in programmed synaptic elimination during adolescence?', *Journal of Psychiatric Research*, vol. 17, no. 4, 1982–3, pp. 319–34.

p. 158, 'brain tissue of adults with schizophrenia': L. J. Garey, W. Y. Ong, T. S. Patel, M. Kanani, A. Davis, A. M. Mortimer, T. R. Barnes and S. R. Hirsch, 'Reduced dendritic spine density on cerebral cortical pyramidal neurons in schizophrenia', *Journal of Neurology, Neurosurgery and Psychiatry*, vol. 65, no. 4, 1998, pp. 446–53.

p. 158, 'lower in young adults with schizophrenia': K. Yuan, W. Qin, G. Wang, F. Zeng, L. Zhao. X. Yang et al., 'Microstructure abnormalities in adolescents with internet addiction disorder', *PLOS ONE*, vol. 6, no. 6, 2011, e20708, doi:10.1371/journal.pone.0020708.

p. 159, 'at high risk of schizophrenia but who do not develop the condition': T. B. Ziermans, P. F. Schothorst, H. G. Schnack, P. C. Koolschijn, R. S. Kahn, J. van Engeland et al., 'Progressive structural brain changes during development of psychosis', *Schizophrenia Bulletin*, vol. 38, no. 3, 2012, pp. 519–30.

p. 160, 'the developmental trajectory of brain structure during adolescence': P. Shaw, D. Greenstein, J. P. Lerch, L. Clasen, R. Lenroot, N. Gogtay, A. Evans, J. Rapoport and J. N. Giedd, 'Intellectual ability and cortical development in children and adolescents', *Nature*, vol. 440, 2006, pp. 676–9.

p. 160, 'developmental changes in cortical thickness were related to IQ': H. G. Schnack, N. E. M. van Haren, R. M. Brouwer, A. Evans, S. Durston, D. L. Boomsma, R. S. Kahn, H. E. Hulshoff Pol, 'Changes in thickness and surface area of the human cortex and their relationship with intelligence', *Cerebral Cortex*, vol. 25, no. 6, 2014, pp. 1608–17.

p. 160, 'depression is the one with the largest impact': World Health Organization, *Global burden of disease*, 2011. See http://www.who.int/topics/global_burden_of_disease/en/.

p. 161, 'tasks with an emotional component': R. Kerestes, C. G. Davey, K. Stephanou, S. Whittle and B. J. Harrison, 'Functional brain imaging studies of youth depression: a systematic review', *NeuroImage: Clinical*, vol. 4, 2013, pp. 209–31.

p. 162, '*The stressed sex*': D. Freeman and J. Freeman, *The stressed sex: uncovering the truth about men, women, and mental health* (Oxford: Oxford University Press, 2015).

p. 163, 'gender differences in puberty onset': J. N. Giedd, J. Blumenthal, N. O. Jeffries, F. X. Castellanos, H. Lui, Z. Zijdenbos, T. Paus, A. C. Evans and J. L. Rapoport, 'Brain development during childhood and adolescence: a longitudinal MRI study', *Nature Neuroscience*, vol. 2, no. 10, 1999, pp. 861–3; R. K. Lenroot, N. Gogtay, D. K. Greenstein, E. M. Wells, G. L. Wallace, L. S. Clasen, J. D. Blumenthal, J. Lerch, A. P. Zijdenbos, A. C. Evans, P. M. Thompson and J. N. Giedd, 'Sexual dimorphism of brain developmental trajectories during childhood and adolescence', *NeuroImage*, vol. 36, no. 4, 2007, pp. 1065–73.

p. 163, 'eighteen months to two years later than girls': B. Bordini and R. L. Rosenfield, 'Normal pubertal development, part I: The endocrine basis of puberty', *Pediatrics Review*, vol. 32, no. 6, 2011, pp. 223–9; S. S. Sun, C. M. Schubert, W. C. Chumlea, A. F. Roche, H. E. Kulin, P. A. Lee, J. H. Himes and A. S. Ryan, 'National estimates of the timing of sexual maturation and racial differences among US children', *Pediatrics*, vol. 110, no. 5, 2002, pp. 911–19.

p. 163, 'largely negated the gender differences': K. L. Mills, A.-L. Goddings, M. M. Herting, R. Meuwese, S.-J. Blakemore, E. A. Crone, R. E. Dahl, B. Güroğlu, A. Raznahan, E. R. Sowell and C. K. Tamnes, 'Structural brain development between childhood and adulthood: convergence across four longitudinal samples', *NeuroImage*, vol. 141, 2016, pp. 273–81.

p. 164, 'Another environmental risk factor': L. Arseneault, M. Cannon, R.

Poulton, R. Murray, A. Caspi and T. E. Moffitt, 'Cannabis use in adolescence and risk for adult psychosis: longitudinal prospective study', *British Medical Journal*, vol. 325, no. 7374, 2002, pp. 1212–13.

p. 165, 'cannabis use in the teenage years': M. H. Meier, A. Caspi, A. Ambler, H. Harrington, R. Houts, R. S. Keefe, K. McDonald, A. Ward, R. Poulton and T. E. Moffitt, 'Persistent cannabis users show neuropsychological decline from childhood to midlife', *Proceedings of the National Academy of Sciences of the United States*, vol. 109, no. 40, 2012, pp. E2657–E2664.

p. 167, 'Binge-drinking is usually defined': J. Jacobus, L. M. Squeglia, S. Bava and S. F. Tapert, 'White matter characterization of adolescent binge drinking with and without co-occurring marijuana use: a 3-year investigation', *Psychiatry Research: Neuroimaging*, vol. 214, no. 3, 2013, pp. 374–81.

p. 167, 'Over half of 15–18-year-olds in the UK': C. J. Armitage, 'Patterns of excess alcohol consumption among school children in two English comprehensive schools', *International Journal of Drug Policy*, vol. 24, no. 5, 2013, pp. 439–44; C. Healey, A. Rahman, M. Faizal and P. Kinderman, 'Underage drinking in the UK: changing trends, impact and interventions. A rapid evidence synthesis', *International Journal of Drug Policy*, vol. 25, no. 1, 2014, pp. 124–32.

p. 167, 'up to a third in the US': H. Wechsler, A. Davenport, G. Dowdall, B. Moeykens and S. Castillo, 'Health and behavioral consequences of binge drinking in college: a national survey of students at 140 campuses', *Journal of the American Medical Association*, vol. 272, no. 21, 1994, pp. 1672–7.

p. 167, 'interfering with school or work': American Psychiatric Association, *Diagnostic and statistical manual of mental disorders*, 5th edn (Lake St Louis, Mo., 2013).

p. 167, 'negative consequences for the developing brain': L. P. Spear, 'Adolescents and alcohol: acute sensitivities, enhanced intake, and later consequences', *Neurotoxicology and Teratology*, vol. 41, Jan.–Feb. 2014, pp. 51–9.

p. 168, 'two or three times as much alcohol as adult rats': S. C. Brunell and L. P. Spear, 'Effect of stress on the voluntary intake of a sweetened ethanol solution in pair-housed adolescent and adult rats', *Alcoholism: Clinical and Experimental Research*, vol. 29, no. 9, pp. 1641–53.

p. 168, 'summarize the findings in a review': S. W. Feldstein-Ewing, A Sakhardande and S.-J. Blakemore, 'The effect of alcohol consumption on the adolescent brain: a systematic review of MRI and fMRI studies of alcohol-using youth', *NeuroImage: Clinical*, vol. 5, 2014, pp. 420–37.

p. 169, 'difficulty in resisting the temptation': M. Pascual, A. Pla, J. Miñarro and C. Guerri, 'Neuroimmune activation and myelin changes in adolescent rats exposed to high-dose alcohol and associated cognitive dysfunction: a review with reference to human adolescent drinking', *Alcohol and Alcoholism*, vol. 49, no. 2, 2014, pp. 187–92.

p. 169, '*greater* engagement of regions involved': R. R. Wetherill, N. Castro, L. M. Squeglia and S. F. Tapert, 'Atypical neural activity during inhibitory processing in substance-naïve youth who later experience alcohol-induced blackouts', *Drug and Alcohol Dependence*, vol. 128, no. 3, 2013, pp. 243–9; R. R. Wetherill, L. M. Squeglia, T. T. Yang and S. S. Tapert, 'A longitudinal examination of adolescent response inhibition: neural differences before and after the initiation of heavy drinking', *Psychopharmacology*, vol. 230, no. 4, 2013, pp. 663–71.

p. 171, 'reduce or abstain from alcohol use': K. M. Lisdahl, E. R. Gilbart, N. E. Wright and S. Shollenbarger, 'Dare to delay? The impacts of adolescent alcohol and marijuana use onset on cognition, brain structure, and function', *Frontiers in Psychiatry*, vol. 4, 2013, article 53; M. A. Monnig, J. S. Tonigan, R. A. Yeo, R. J. Thoma and B. S. McCrady, 'White matter volume in alcohol use disorders: a meta-analysis', *Addiction Biology*, vol. 18, no. 3, 2013, pp. 581–92; K. A. Welch, A. Carson and S. M. Lawrie, 'Brain structure in adolescents and young adults with alcohol problems: systematic review of imaging studies', *Alcohol and Alcoholism*, vol. 48, no. 4, 2013, pp. 433–44.

p. 171, 'effects of screens on brain development': https://www.theguardian.com/science/head-quarters/2017/jan/06/screen-time-guidelines-need-to-be-built-on-evidence-not-hype

p. 171, 'The research simply has not been done': K. L. Mills, 'Effects of internet use on the adolescent brain: despite popular claims, experimental evidence remains scarce', *Trends in Cognitive Science*, vol. 18, no. 8, 2014, pp. 385–7.

p. 172, 'for this discovery of yours': Plato, *Phaedrus* (360 BCE), 274c–275b.

p. 172, '"confusing and harmful" to the mind': C. Gessner, *Bibliotheca universalis, sive catalogus omnium scriptorum locupletissimus, in tribus linguis, Latina, Graeca, & Hebraica: extantium & non extantium veterum & recentiorum* (1545).

p. 173, 'analysed reports on screen-time and mental health': A. Przybylski and N. Weinstein, 'A large-scale test of the Goldilocks hypothesis: quantifying the relations between digital-screen use and the mental well-being of adolescents', *Psychological Science*, vol. 28, no. 2, 2017, pp. 204–15.

11: Educating the brain

p. 177, 'have written a book about neuroscience': S.-J. Blakemore and U. Frith, *The learning brain: lessons for education* (Oxford: Blackwell, 2005).

p. 177, 'a "bridge too far"': J. T. Bruer, 'Education and the brain: a bridge too far', *Educational Researcher*, vol. 26, no. 8, 1997, pp. 4–16.

p. 180, 'could not be replicated': Open Science Collaboration, 'Estimating the reproducibility of psychological science', *Science*, vol. 349, no. 6251, 2015, aac4716-6.

p. 180, 'early environments are crucial to brain development': D. Wastell and S. White, 'Blinded by neuroscience: social policy, the family and the infant brain', *Families, Relationships and Societies*, vol. 1, no. 3, 2012, pp. 397–414.

p. 181, 'have worked on this subject for many years': E. J. McCrory, M. I. Gerin and E. Viding, 'Annual research review: childhood maltreatment, latent vulnerability and the shift to preventative psychiatry – the contribution of functional brain imaging', *Journal of Child Psychology and Psychiatry*, vol. 58, no. 4, April 2017, pp. 338–57, doi: 10.1111/jcpp.12713.

p. 181, 'specific and much more subtle differences': L. Lim, J. Radua and K. Rubia, 'Gray matter abnormalities in childhood maltreatment: a voxel-wise meta-analysis', *American Journal of Psychiatry*, vol. 171, no. 8, 2014, pp. 854–63.

p. 181, '"We've known about these outcomes"': P. Butler, 'Policymakers seduced by neuroscience to justify early intervention agenda', *Guardian*, 6 May 2014, https://www.theguardian.com/society/2014/may/06/policymakers-neuroscience-justify-early-intervention-agenda-bruce-perry.

p. 182, 'with or without neuroscience terminology': D. S. Weisberg, F. C. Keil, J. Goodstein, E. Rawson and J. R. Gray, 'The seductive allure of neuroscience explanations', *Journal of Cognitive Neuroscience*, vol. 30, no. 3, 2008, pp. 470–7.

p. 184, 'writes prolifically about "bad science"': B. Goldacre, *Bad science: quacks, hacks, and big pharma flacks* (London: Faber, 2010).

p. 184, 'randomized controlled trials in schools': B. Goldacre, 'Building evidence into education', March 2013, http://media.education.gov.uk/assets/files/pdf/b/ben%20goldacre%20paper.pdf.

p. 185, 'improve different types of working memory': www.cogmed.com.

p. 185, 'In one game': http://www.spaceminespatrol.com/.

pp. 185–6, 'played working memory video-games': T. Klingberg, H. Forssberg and H. Westerberg, 'Training of working memory in children with ADHD', *Journal of Clinical and Experimental Neuropsychology*, vol. 24, no. 6, Sept. 2002, pp. 781–91.

p. 186, 'other domains, including arithmetic': S. Bergman-Nutley and T. Klingberg, 'Effect of working memory training on working memory, arithmetic and following instructions', *Psychological Research*, vol. 78, no. 6, 2014, pp. 869–77, doi:10.1007/s00426-014-0614-0.

p. 186, 'the more their non-verbal reasoning improved': S. M. Jaeggi, M. Buschkuehl, J. Jonides and W. J. Perrig, 'Improving fluid intelligence with training on working memory', *Proceedings of the National Academy of Sciences of the United States*, vol. 105, no. 19, 2008, pp. 6829–33.

p. 187, 'improves anything except working memory': T. S. Redick, Z. Shipstead, T. L. Harrison, K. L. Hicks, D. E. Fried, D. Z. Hambrick, M. J. Kane and R. W. Engle, 'No evidence of intelligence improvement after working memory training: a randomized, placebo-controlled study', *Journal of Experimental Psychology: General*, vol. 142, no. 2, May 2013, pp. 359–79, doi: 10.1037/a0029082.

p. 187, 'including working memory and reasoning': A. M. Owen, A. Hampshire, J. A. Grahn, R. Stenton, S. Dajani, A. S. Burns and C. G. Ballard, 'Putting brain training to the test', *Nature*, vol. 465, 2010, pp. 775–8, doi:10.1038/nature09042.

p. 187, 'it's unlikely that your general IQ will increase': D. J. Simons, W. R. Boot, N. Charness, S. E. Gathercole, C. F. Chabris, D. Z. Hambrick and E. A. Stine-Morrow, 'Do "brain-training" programs work?', *Psychological Science in the Public Interest*, vol. 17, no. 3, 2016, pp. 103–86.

p. 188, 'UNICEF estimates': see https://data.unicef.org/topic/education/secondary-education/; https://www.unicef.org/factoftheweek/index_50244.html.

p. 189, 'sleep deprivation has a detrimental effect on learning': M. Walker, *Why we sleep* (London: Allen Lane, 2017).

p. 190, 'One explanation, put forward': R. Foster, 'Why teenagers really do need an extra hour in bed', *New Scientist*, 17 April 2013, https://www.newscientist.com/article/mg21829130-100-why-teenagers-really-do-need-an-extra-hour-in-bed/; Mary A. Carskadon, 'Sleep in adolescents: the perfect storm', *Pediatric Clinics of North America*, vol. 58, no. 3, 2011, pp. 637–47.

p. 191, 'Steinberg makes a clear case': L. Steinberg, 'Should the science of adolescent brain development inform public policy?', *American Psychologist*, vol. 64, no. 8, 2009, p. 739; E. Cauffman and L. Steinberg, 'Researching adolescents' judgment and culpability', in T. Grisso and R. G. Schwartz (eds), *Youth on trial: a developmental perspective on juvenile justice* (Chicago: University of Chicago Press, 2000), pp. 325–43; L. Steinberg and E. S. Scott, 'Less guilty by reason of adolescence: developmental immaturity, diminished responsibility, and the juvenile death penalty', *American Psychologist*, vol. 58, no. 12, 2003, pp. 1009–18.

pp. 191–2, 'no matter how serious their crime': see M. Beckman, 'Crime, culpability and the adolescent brain', *Science*, vol. 305, 30 July 2004, https://deathpenaltyinfo.org/node/1225.

p. 192, 'discussed at length by Terrie Moffitt': T. E. Moffitt, 'Adolescence-limited and life-course-persistent antisocial behavior: a developmental taxonomy', *Psychological Review*, vol. 100, no. 4, 1993, pp. 674–701.

12: It's the journey that matters

pp. 194–5, 'the development of divergent thinking': S. W. Kleibeuker, C. K. W. De Dreu and E. A. Crone, 'The development of creative cognition across adolescence:

distinct trajectories for insight and divergent thinking', *Developmental Science*, vol. 16, no. 1, pp. 2–12.

p. 198, 'a short extract from the play': From the play *Brainstorm*, by Ned Glasier, Emily Lim and Company Three, produced at the Park Theatre, National Theatre and on BBC iPlayer as part of Battersea Arts Centre's Live from Television Centre, 2015–16: www.companythree.co.uk/brainstorm. *Brainstorm* is available as a playtext published by Nick Hern Books, with a guide for any group of young people who want to make their own version.

Acknowledgements

Many friends, family members, colleagues and young people have supported me throughout the process of planning and writing this book. Every chapter was subjected to the sharp editorial eyes of Kathleen Ball and Lucy Foulkes. Their suggestions, tweaks and comments have been invaluable. I am grateful to Emily Garrett, Cait Griffin and Jovita Leung, who helped with fact-checking and copy-editing. Tim Dalgleish, Pete Etchells, Anne-Lise Goddings, James Kilner and Kate Mills, as well as my parents, Colin and Andree Blakemore, read parts of my book and their feedback has been incredibly useful and gratefully received.

Each year we host school students in our lab for work experience, and I asked our work experience students in recent years to read and give me feedback on various chapters. Thank you to Daisy Calder, Rebecca Delaney, Yasmin Farooqui Caliz, Harriet Gould, Natalya Helvadjian and Hannah Knight for their comments. Susanne Malik helped with some of the research.

I am proud to have worked at University College London for many years and I appreciate the constant support of my university. I gratefully acknowledge the funding that has supported my research over the past two decades, including grants from the European Research Council, the Leverhulme Foundation, the Nuffield Foundation, the Royal Society, the Wellcome, and the Jacobs Foundation, which honoured me with the Klaus J. Jacobs Prize in 2015.

The research I describe in this book would not have been possible without the encouragement of my mentors and the dedicated hard work, intellectual drive and creative thinking of my research group. There are too many to list them all here, but I'd like to mention the following researchers who have spent time in my lab and have contributed enormously to the research described in this book. Guillaume Barbalat, Kathrin Cohen Kadosh, Iroise Dumontheil, Lucy Foulkes, Lisa Knoll, Lara Menzies and Stefano Palminteri were postdocs in my group. Jack Andrews, Zillah Boraston, Stephanie Burnett-Heyes, Suparna Choudhury, Jennifer Cook, Delia Fuhrmann, Anne-Lise Goddings, Hauke Hillebrandt, Emma Kilford, Kate Mills, Cat Sebastian, Stephanie Thompson and Laura Wolf were, or are, PhD students under my supervision. Narges Bazargani, Hanneke den Ouden, Hugo Fleming, Emily Garrett, Annie Gaule, Cait Griffin, Rachael Houlton, Elina Jacobs, Sarah Jensen, Eduard Klapwijk, Olivia Küster, Jovita Leung, Lucia Magis-Weinberg, Alex Moscicki, Ashok Sakhardande, Maximilian Scheuplein and Leonora Weil were, or are, research assistants or affiliate students.

Scientific research is a collaborative endeavour. I am lucky to have worked with many people from around the world, each of whom has played an instrumental role in my research on adolescence. Conversations with many collaborators have helped to shape this book. In particular, I'd like to thank Nick Allen, Ian Apperly, Vaughan Bell, Geoff Bird, Silvia Bunge, Tony Charman, Eveline Crone, Ron Dahl, Tim Dalgleish, Sarah Feldstein-Ewing, Chris Frith, Uta Frith, Jay Giedd, Paul Howard-Jones, Willem Kuyken, Eamon McCrory, Mairead MacSweeney, Jennifer Pfeifer, Danny Pine, Nichola Raihani, Jon Roiser, Susanne Schweizer, Sophie Scott, Tali Sharot, Sukhi Shergill, Maarten Speekenbrink, Laurence Steinberg, Christian Tamnes, Anne-Laura van Harmelen, Essi Viding, Russell Viner, Vincent Walsh and Mark Williams. Working with these people, in many cases over several years or even decades, makes doing science extremely rewarding.

My agent, Will Francis, has been unwaveringly supportive from the start of this project. He has a wisdom and knowledge about books that is inspiring. I'm grateful that Transworld saw the potential of my book, and Susanna Wadeson, my editor, has been extraordinarily helpful and encouraging throughout the writing process. Her expertise and experience know no bounds. Gillian Somerscales' copy-editing has been instrumental. These people have inspired me with their talent and expertise.

Finally, I am lucky to have such supportive parents, sisters, children and family, who have given me their unbridled encouragement, as have my friends, including friends I made during my own teenage years.

Illustration sources

p. 4, 'Graph showing that adolescent mice spend more time drinking alcohol when with their cage-mates': redrawn by Liane Payne from S. Logue, J. Chein, T. Gould, E. Holliday and L. Steinberg, 'Adolescent mice, unlike adults, consume more alcohol in the presence of peers than alone', *Developmental Science*, vol. 17, no. 1, 2014, pp. 79–85.

p. 12, 'The average age range for the onset of several different psychological disorders': redrawn by Liane Payne from R. C. Kessler, P. Berglund, O. Demler, R. Jin, K. R. Merikangas and E. E. Walters, 'Lifetime prevalence and age-of-onset distributions of DSM-IV disorders in the National Comorbidity Survey Replication', *Archives of General Psychiatry*, vol. 62, 2005, pp. 593–602. Data compiled by Delia Fuhrmann.

p. 33, 'The Stoplight task': redrawn by Liane Payne from M. Gardner and L. Steinberg, 'Peer influence on risk taking, risk preference, and risky decision making in adolescence and adulthood: an experimental study', *Developmental Psychology*, vol. 41, no. 4, 2005, pp. 625–35.

p. 41, 'The see-saw of decision-making': K. L. Mills, A.-L. Goddings and S.-J. Blakemore, 'Drama in the teenage brain', *Frontiers for Young Minds*, vol. 2, no. 16, 2014, pp. 1–5.

p. 47, 'The social influence effect in risk perception': L. J. Knoll, L. Magis-Weinberg, M. Speekenbrink and S.-J. Blakemore, 'Social influence of risk perception during adolescence', *Psychological Science*, 25, 2015, pp. 583–92.

p. 51, 'The human brain': Getty Images.

p. 53, 'The brain's lateral surface', 'The brain's medial surface': Liane Payne.

p. 57, 'Phineas Gage': Ruth Murray.

p. 60, 'Detail of a synapse': Liane Payne.

p. 69, 'An MRI scanner': Photo © Tali Sharot.

p. 81, 'The development of grey matter and white matter during adolescence':

redrawn by Liane Payne from K. L. Mills, A.-K. Goddings, M. M. Herting, R. Meuwese, S.-J. Blakemore, E. A. Crone, R. E. Dahl, B. Güroğlu, A. Raznahan, E. R. Sowell and C. K. Tamnes, 'Structural brain development between childhood and adulthood: convergence across four longitudinal samples', *NeuroImage*, vol. 141, 2016, pp. 273–81; C. K. Tamnes, M. M. Herting, A. L. Goddings, R. Meuwese, S.-J. Blakemore, R. E. Dahl, B. Güroğlu, A. Raznahan, E. R. Sowell, E. A. Crone and K. L. Mills, 'Development of the cerebral cortex across adolescence: a multisample study of interrelated longitudinal changes in cortical volume, surface area and thickness', *Journal of Neuroscience*, vol. 27, no. 12, 2017, pp. 3402–12.

p. 82, 'Detail of a neuron': Liane Payne.

p. 89, 'The Shopping Task': redrawn by Liane Payne from T. Shallice and P. W. Burgess, 'Deficits in strategy application following frontal lobe damage in man', *Brain*, vol. 114, no. 2, 1991, pp. 727–41.

p. 93, 'An example of the non-verbal reasoning task': Ruth Murray.

p. 105, 'An example of the stimuli in the Shapes Task': redrawn by Ruth Murray from I. Dumontheil, R. Houlton, K. Christoff and S.-J. Blakemore, 'Development of relational reasoning during adolescence', *Developmental Science*, 13, 2010, pp. F15–F24.

p. 112, 'The Director Task': redrawn by Ruth Murray from I. Dumontheil, I. A. Apperly and S.-J. Blakemore, 'Online usage of theory of mind continues to develop in late adolescence', *Developmental Science*, vol. 13, no. 2, 2010, pp. 331–8.

p. 122, 'Examples of cartoons': redrawn by Ruth Murray from H. L. Gallagher, F. Happé, N. Brunswick, P. C. Fletcher, U. Frith and C. D. Frith, 'Reading the mind in cartoons and stories: an fMRI study of "Theory of Mind" in verbal and nonverbal tasks', *Neuropsychologia*, vol. 38, no. 1, 2000, pp. 11–21.

p. 124, 'Sincere or ironic?': redrawn by Ruth Murray from A. T. Wang, S. S. Lee, M. Sigman and M. Dapretto, 'Developmental changes in the neural basis of interpreting communicative intent', *Social Cognitive and Affective Neuroscience*, vol. 1, no. 2, 2006, pp. 107–21.

p. 138, 'Grey matter development in three brain regions': redrawn by Liane Payne from K. L. Mills, A.-L. Goddings, L. S. Clasen, J. N. Giedd and S.-J. Blakemore, 'The developmental mismatch in structural brain maturation during adolescence', *Developmental Neuroscience*, vol. 36, nos 3–4, 2014, pp. 147–60.

p. 146, 'The gambling task': redrawn by Liane Payne from S. Burnett, N. Bault, G. Coricelli and S.-J. Blakemore, 'Adolescents' heightened risk-seeking in a probabilistic gambling task', *Cognitive Development*, vol. 25, no. 2, 2010, pp. 183–96.

p. 149, 'Resisting temptation': Ruth Murray.

p. 152: 'Self-control in childhood and outcomes in adulthood': redrawn by Liane Payne from Ed Yong, 'Self-control in childhood predicts health and wealth in adulthood', http://phenomena.nationalgeographic.com/2011/01/24/self-control-in-childhood-predicts-health-and-wealth-in-adulthood/.

p. 173, 'Well-being and screen-time among 15-year-olds': redrawn by Liane Payne from A. Przybylski and N. Weinstein, 'A large-scale test of the Goldilocks hypothesis: quantifying the relations between digital-screen use and the mental well-being of adolescents', *Psychological Science*, vol. 28, no. 2, 2017, pp. 204–15.

Index

page numbers in *italics* refer to pages with illustrations

Sarah-Jayne Blakemore is a professor in cognitive neuroscience at University College London. She has published over 100 papers in scientific journals, and won multiple major awards for her research, including the BPS Spearman Medal 2006, the Turin Young Mind & Brain Prize 2013, the Royal Society Rosalind Franklin Award 2013, and the Klaus J. Jacobs Research Prize 2015. She was named in the *Times* Young Female Power List 2014 and was one of only four scientists on the *Sunday Times* 100 Makers of the 21st Century 2014.

BB S

PublicAffairs is a publishing house founded in 1997. It is a tribute to the standards, values, and flair of three persons who have served as mentors to countless reporters, writers, editors, and book people of all kinds, including me.

I. F. Stone, proprietor of *I. F. Stone's Weekly*, combined a commitment to the First Amendment with entrepreneurial zeal and reporting skill and became one of the great independent journalists in American history. At the age of eighty, Izzy published *The Trial of Socrates*, which was a national bestseller. He wrote the book after he taught himself ancient Greek.

Benjamin C. Bradlee was for nearly thirty years the charismatic editorial leader of *The Washington Post*. It was Ben who gave the *Post* the range and courage to pursue such historic issues as Watergate. He supported his reporters with a tenacity that made them fearless and it is no accident that so many became authors of influential, best-selling books.

Robert L. Bernstein, the chief executive of Random House for more than a quarter century, guided one of the nation's premier publishing houses. Bob was personally responsible for many books of political dissent and argument that challenged tyranny around the globe. He is also the founder and longtime chair of Human Rights Watch, one of the most respected human rights organizations in the world.

• • •

For fifty years, the banner of Public Affairs Press was carried by its owner Morris B. Schnapper, who published Gandhi, Nasser, Toynbee, Truman, and about 1,500 other authors. In 1983, Schnapper was described by *The Washington Post* as "a redoubtable gadfly." His legacy will endure in the books to come.

Peter Osnos, *Founder*